Development of Marine Resources

From the ***Seas and Oceans*** Set
coordinated by
André Mariotti and Jean-Charles Pomerol

Development of Marine Resources

Edited by

André Monaco
Patrick Prouzet

ISTE

WILEY

First published 2014 in Great Britain and the United States by ISTE Ltd and John Wiley & Sons, Inc.

ISTE Ltd
27-37 St George's Road
London SW19 4EU
UK

www.iste.co.uk

John Wiley & Sons, Inc.
111 River Street
Hoboken, NJ 07030
USA

www.wiley.com

Library of Congress Control Number: 2014953026

British Library Cataloguing-in-Publication Data
A CIP record for this book is available from the British Library
ISBN 978-1-84821-705-8

Contents

Foreword

We have been asked by ISTE to stimulate work in the area of the environment. Therefore, we are proud to present the "Seas and Oceans" set of books, edited by André Monaco and Patrick Prouzet.

Both the content and the organization of this collection have largely been inspired by the reflection, initiatives and prospective works of a wide variety of national, European and international organizations in the field of the environment.

The "oceanographic" community, in France and internationally – which is recognized for the academic quality of the work it produces, and is determined that its research should be founded on a solid effort in the area of training and knowledge dissemination – was quick to respond to our call, and now offers this set of books, compiled under the skilled supervision of the two editing authors.

Within this community, there is a consensus about the need to promote an interdisciplinary "science of systems" – specifically in reference to the Earth's own "system" – in an all-encompassing approach, with the aim of providing answers about the planet's state, the way it works and the threats it faces, before going on to construct scenarios and lay down the elementary foundations needed for long-term, sustainable environment management, and for societies to adapt as required. This approach facilitates the shift of attention from this fundamental science of systems (based on the analysis of the processes at play, and the way in which they interact at all levels and between all the constituent parts making up the global system) to a "public" type of science, which is finalizable and participative, open to decision-makers, managers and all those who are interested in the future of our planet.

In this community, terms such as "vulnerability", "adaptation" and "sustainability" are commonly employed. We speak of various concepts, approaches or technologies, such as the value of ecosystems, heritage, "green" technologies, "blue" chemistry and renewable energies. Another foray into the field of civilian science lies in the adaptation of research to scales which are compatible with the societal, economic and legal issues, from global to regional to local.

All these aspects contribute to an in-depth understanding of the concept of an ecosystemic approach, the aim of which is the sustainable usage of natural resources, without affecting the quality, the structure or the function of the ecosystems involved. This concept is akin to the "socio-ecosystem approach" as defined by the Millennium Assessment (http://millenniumassessment.org).

In this context, where the complexity of natural systems is compounded with the complexity of societies, it has been difficult (if only because of how specialized the experts are in fairly reduced fields) to take into account the whole of the terrestrial system. Hence, in this editorial domain, the works in the "Seas and Oceans" set are limited to fluid envelopes and their interfaces. In that context, "sea" must be understood in the generic sense, as a general definition of bodies of salt water, as an environment. This includes epicontinental seas, semi-enclosed seas, enclosed seas, or coastal lakes, all of which are home to significant biodiversity and are highly susceptible to environmental impacts. "Ocean", on the other hand, denotes the environmental system, which has a crucial impact on the physical and biological operation of the terrestrial system – particularly in terms of climate regulation, but also in terms of the enormous reservoir of resources they constitute, covering 71% of the planet's surface, with a volume of 1,370 million km^3 of water.

This set of books covers all of these areas, examined from various aspects by specialists in the field: biological, physical or chemical function, biodiversity, vulnerability to climatic impacts, various uses, etc. The systemic approach and the emphasis placed on the available resources will guide readers to aspects of value-creation, governance and public policy. The long-term observation techniques used, new techniques and modeling are also taken into account; they are indispensable tools for the understanding of the dynamics and the integral functioning of the systems.

Finally, treatises will be included which are devoted to methodological or technical aspects.

The project thus conceived has been well received by numerous scientists renowned for their expertise. They belong to a wide variety of French national and international organizations, focusing on the environment.

These experts deserve our heartfelt thanks for committing to this effort in terms of putting their knowledge across and making it accessible, thus providing current students with the fundaments of knowledge which will help open the door to the broad range of careers that the area of the environment holds. These books are also addressed to a wider audience, including local or national governors, players in the decision-making authorities, or indeed “ordinary” citizens looking to be informed by the most authoritative sources.

Our warmest thanks go to André Monaco and Patrick Prouzet for their devotion and perseverance in service of the success of this enterprise.

Finally, we must thank the CNRS and Ifremer for the interest they have shown in this collection and for their financial aid, and we are very grateful to the numerous universities and other organizations which, through their researchers and engineers, have made the results of their reflections and activities available to this instructional corpus.

André MARIOTTI
Professor Emeritus at University Pierre and Marie Curie
Honorary Member of the Institut Universitaire de France
France

Jean-Charles POMEROL
Professor Emeritus at University Pierre and Marie Curie
France

1

Fishing in the Mediterranean, Past and Present: History and Technical Changes

The historical context of the maritime sector can present certain difficulties for historians and fishery specialists. Seas and oceans are mostly worlds without archives, often mistakenly thought of as immutable. Most often glossed over by biologists, more than ever does it seem necessary to ask questions about the past of the underwater world. It must today be at the heart of all reflections that condition the defining of policies for the management of fisheries. While the challenges faced by Mediterranean fishing, climate change, acidification and the need for a rigorous management of stock, are no different from those found in all seas of the world, they are all the more applicable in this closed basin and must engage the responsibilities of all the surrounding states in the 21st Century. Dedicated for millennia to fishing and the movement of men and goods, this "liquid continent with a solid border", as it was called in the 1930s by the poet Gabriel Audisio and his friends from the *Cahiers du Sud*, has witnessed the rise of new uses in the last 50 years, which present important problems for fisheries. The rise of leisure fishing, harm to coastal fishing caused by the damaging of marine ecosystems and the need to provide for urban and tourist markets increasingly demand of sea products today, influence the maintenance

---

Chapter written by Daniel FAGET and Jacques SACCHI.

or the survival of professional fishing in all of the basin. However, the long history of Mediterranean fishing, far from unmoving, shows that the men and women who formed it over centuries were able to adapt their methods to the fluctuating conditions of access and exploitation of fishing resources. The future of fishing in the Mediterranean requires mastering the effects of industrialization and urbanization on ecosystems, also must take into account the cultural aspects, including traditional expertise, for a better management of this medium. Compiling a comprehensive history of fishing would appear unfeasible in the confines of this chapter. A choice has therefore been made first to present a reflection on historical methods of fishing. A table of traditional techniques, the organization of communities and their plurality is envisaged here (section 1.1). Second, the successive changes that have transformed this activity sector since the modern era are analyzed (sections 1.2 and 1.3). The third component of this approach proposes some reflection on the characteristic of Mediterranean fishing at the start of the 21st Century. This last part will be a description of the current flotillas in operation, while looking at the challenges faced by this sector of activity: environmental changes, changes in how people consume, the redefining of the fishing boss, between collective needs for the protection of the habitat and the need to fulfill the demands of the market (section 1.4).

1.1. Mediterranean fishing of the past (18th Century)

Structured by religious brotherhoods and communal efforts, fishing communities efficiently control the exploitation of the natural medium in the modern era. The large variety of fishing techniques used, as well as the characteristics of an "Ancient Regime" style of consumption, marked by a chronic lack of protein, explains the extreme diversity of the products offered at the fishmonger's stall.

1.1.1. *Brotherhoods and prud'homies: old forms of regulation of the fishing world*

Being present as early as the 15th Century on the north coast of the Mediterranean, the brotherhoods, Italian *confraternite*, Catalan *gremis* or the Spanish *cofradias*, appeared as the dominating form of organization in the fishing community. Placed under the protection of the Virgin Mary or a patron saint, Saint Peter, Saint Elme or Saint Roch, the brotherhoods were first of all religious structures that not only guaranteed their members' collective solidarity in the case of an accident (loss of ships or fishing materials), but also looked after the souls of the dead through prayers and processions. They were also professional regulatory structures. Made up of all the fishing bosses, i.e. the boat owners, they enacted precise fishing regulations, most often passed on orally. The distribution of fishing zones (the Provençale "postes"), the mesh size of the nets, the size of the hooks, the quality of the baits and a strict calendar fixed by the community [FAG 11], thus precisely defined the modes of exploitation of a "fishing ground" [FER 01] whose spatial limits are strictly delimited. The names used by the fishing communities of the past clearly illustrate this distribution of the marine space, which was split into microterritories: for example the two seas of *Amoun* and *Avau*, which split the organization of fishing in Marseilles in the 17th Century [FAG 11]. Exceptionally, the organization of the community would depend on two structures: the *prud'hommes*, a tribunal made up of representatives of the profession, are clearly distinguished from the brotherhood, which would in this case be reduced to a spiritual function and a function of assistance. The only jurisdiction in all of the ports of the Mediterranean, the corporation of *prud'hommes* of Marseilles would appear today as the most accomplished form of self-regulatory organism for fishing activities [BER 98]. It benefits from its age, since the municipal authorities allow the community as early as the 14th Century to choose its own *probi homines*, its wise men, to sort out any conflicts related to fishing. There are four of them, renewed each year and elected by a simple vote by show of hands. These Marseilles *prud'hommes* provide public justice, orally, freely and without the possibility of appeal. A tribunal of experts and recognized as such, the Marseilles *prud'hommes* thus avoid the suppression of corporations put in place by the revolutionary laws of the 2nd and 17th March [FAG 11]. A model of professional organization, as early as the last decade of the 18th Century, it became the dominating form of justice within the fishing communities of the French Mediterranean coast.

Figure 1.1. *Provençale fishing prud'hommes (source: Musée Ciotaden)*

COMMENTS ON FIGURE 1.1.– As Mediterranean fishing boss communities, the *prud-hommes* appeared for the first time in Marseilles in 1481, officially recognized by Louis XI's royal charter. Extended to all fishing communities after the French Revolution, the *prud'homme* form of organization found its definitive form in a decree from 1859. The function of the *prud'hommes* is to sort out conflicts between fishermen and regulate the access to fishing zones depending on their jurisdiction. Long neglected by the legal authorities, especially during attempts to develop industrial fishing, they have been, since 1994 by decree of the *Affaires Maritimes*, systematically consulted before any regulation is made in maritime affairs. As a decentralized power of management and authority, the *prud'hommes* constitute a model of management and governance of fisheries ensured by the polyvalence of the activities and making the fishers aware of their responsibility, which are the optimal conditions for the proper exploitation of the resources. "The five *prud'hommes* wore hose, a doublet and a black coat with a white band. On their heads they wore a hat with large edges. Their faces were tanned, and they represented the elite of the maritime population of the town and of the gulf" [SUE 45].

1.1.2. *Plural communities*

Present over the entire Mediterranean coast, traditional fishing communities offer much diversity. The differences first concern the forms of habitat and insertion in the coastal space. From the simple Languedoc or Moroccan rosewood hut [FER 01, PAY 07] to the specific urban quarter, the fisherman's habitat appears as the result of natural conditions (dirtiness of the coast), and also depends on historical processes that can reflect the age of a community (Saint-Jean quarter in Marseilles, Jonquières quarter in Martigues, etc.), or reflect political decisions, often made later (Barceloneta quarter in Barcelona, created from scratch in 1753, La Bordigue quarter in Sète, after the 17th Century) [CAB 95]. On top of these differences in accommodation, a plurality of the activities that are not entirely dedicated to fishing can be added. Better than the classes system etablished by the French Royal administration, the study parish registers also frequently highlights the professional instability of the people of the sea, successively recorded professionally as "brassiers"[1] rather than fisher[2].

Like for the Atlantic coast, the presence of a cultivatable inland explains the coexistence in the Mediterranean, within fishing families, of time dedicated to the cultivation of wine grapes or wheat and periods dedicated to fishing. The availability of agricultural resources, which sometimes transforms income taken from the sea into significant revenue, explains the choice of one type of fishing over another by communities. In the Languedoc and Roussillon, the fisher-winemakers of Leucate, Banyuls or Collioure, masters of the *sardinal*[3] or of small fishing in lagoons, thus opposed the fishing owners of Gruissan or Sète in the 18th Century, who were converted to the *pêche au boeuf*[4] very early on (section 1.2) [LAR 97]. For the most part an opportunist, able to make the most of any positive variations offered by the resource, the traditional fisherman adapts his trade according to the season. He knows how to use the *boguière* or the *thonaire*[5] with the same dexterity as the

1 Labourers.

2 In 1668, Jean-Baptiste Colbert, the minister of Louis XIV, created the class system, which enrolls and identifies all seamen to provide regular service to the Royal Navy.

3 The sardinal is a drifting gillnet on the surface for catching sardines.

4 The "pêche au boeuf" is a pair-trawling technique that consists of towing a trawl simultaneously with two vessels.

5 The *boguière* targets the bogues (*Boops boops*) and the *thonaire* aims to capture bluefin tuna. They are like the drifting sardinal surface gillnets or are set on the seabed.

girelier[6] or the *eissaugue*[7], due to ancestral knowledge passed down from father to son. The use of all types of traditional fishing, more than the limits of his expertise, only depends on the financial capabilities that condition the buying of certain materials, whose cost is often greater than the value of the vessel itself.

1.1.3. *Diversity of traditional techniques in the Mediterranean*

Under the watchful eye of brotherhoods or *prud'hommes*, Mediterranean fishermen use in the modern era techniques that were already known in Antiquity, and often represented in mosaics of the Greco-Roman civilization, an example being the one found in the Villa del Casale in Sicily, dating back to the 3rd Century AD. Passed on through the vernacular, the expertise attached to these techniques is rarely the object of treaties or professional manuals. They are part of an oral culture of apprenticeship, provided on a vessel, aimed at sailors. The materials used can be classified into two categories. The first category is that of static gear, traps, coastal fishing lines, longlines and nets – whose extreme variety reflects the species that are being caught. Among these, we can distinguish bottom gillnets with a single *aumée*[8], the trammel nets superimposing three *aumées* or net panels, aimed at catching benthic fish, gillnets floating on the surface, aimed at catching pelagic or semi-pelagic species (tuna, sardines, anchovies, etc.). This first group of static gears, opposed to all the mobile gear, is essentially composed of towed nets [MAR 05]. Whether manipulated from the coast by hand (Provençale *eissaugue* and Languedoc *boulier*), or from a vessel (*gangui*, Languedoc "*peche a vache*", Albufera or Valence *gànguil*)[9], these nets have the particularity of sweeping the posidonia prairies, the beds of silt or

6 The *girelier* is a fish trap of wicker or iron wire designed for capturing wrasse (*Coris julis*).

7 The *eissaugues* or *issaugues* referred to a type of towed net similar to beach seines and that were dragged from land by several men to catch various pelagic and demersal species mainly living in coastal sea herb beds [GOU 97].

8 An *aumée* is a layer of netting made up of meshes of different sizes according to the type of fishing being practiced.

9 Ancestor to our modern trawlers, in order to maintain its horizontal opening the *gangui* needed to be towed by two vessels (pair-trawling) or to be held up by two yards fixed to the bow and stern of the vessel (*pêche à la vache*).

of coral sands, to find flat fish and elasmobranchii fish. Whether static or mobile, these different fishing gears are made up of fragile materials and are characterized by rapid wear. Their manufacture is supplied by a highly active artisan industry, mostly gone today, and their maintenance calls upon practices found on most of the Mediterranean coast. Nets made of hemp, which retain humidity, are subject to alteration, the damage caused by which delays the process of dying. This is carried out with the help of the bark from the Alep pine tree (*Pinus halepensis*), from which a decoction is obtained by boiling it in the community cauldron. The fish traps and crab traps (*gireliers*, Provençale *emborniers* and Spanish *nansas*) are most often made from myrtle sticks (*Myrtus communis*), their imputrescibility ensuring the longevity of the materials. If, for the most part, the manufacture of almadraba nets uses the same materials as those used in the elaboration of other fishing gear, the dimensions of these fixed fisheries and their maintenance costs are enough to put them in a class apart from that of small-scale artisanal fishing.

Figure 1.2. *Eissaugue (source: etching taken from Duhamel Du Monceau Henri Louis, Traité général des pêches, Guillaume de Bure, Paris, 1782)*

COMMENTS ON FIGURE 1.2.– A dragged net manipulated from land, the beach seine is a piece of fishing gear whose history goes back to Antiquity. Called *eissaugue* in the French Provence, this net required a consistent coastal line, against a shallow infracoastal space free of rocks. This technique, maintained throughout the Middle Ages, preceded the growth from the 17th Century of open sea trawling. It was, however, still in use in the first half of the 20th Century.

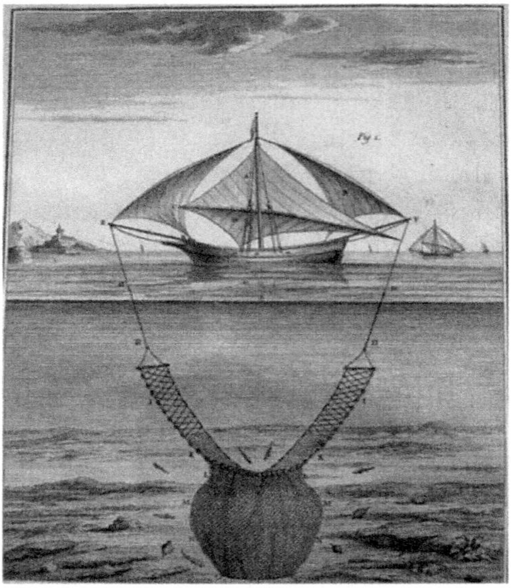

Figure 1.3. *Tartane (source: etching taken from Duhamel Du Monceau Henri Louis, Traité général des pêches, Guillaume de Bure, Paris, 1782)*

COMMENTS ON FIGURE 1.3.– At the end of the 17th Century, a new, powerful, polyvalent vessel appeared in the Mediterranean world. This new vessel was armed with a Latin sail. It was equipped with special nets with high sea fishing. With an important tonnage (between 30 and 50 tons, 1 ton = 2.83m^3), this vessel was first built in Martigues (called *tartana* or *martingana* in 17th Century Italy). Parallel to the distribution of the sardinal, the development of the vessel was at the origin of the development of new activity, dragged fishing techniques at sea.

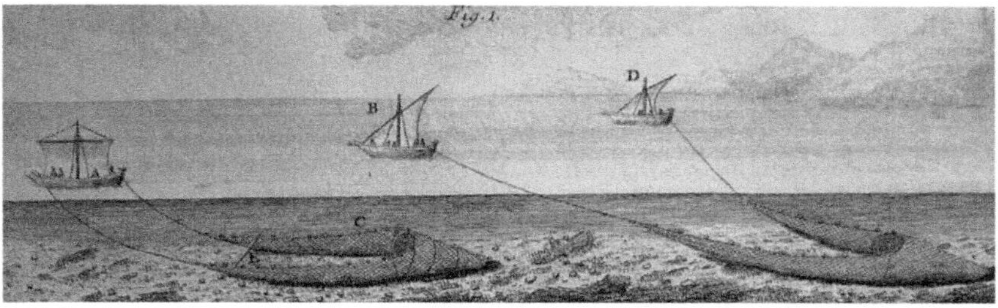

Figure 1.4. *Gangui "à la vache or plow" fishing (source: etching taken from Duhamel Du Monceau Henri Louis, Traité général des pêches, Guillaume de Bure, Paris, 1782)*

COMMENTS ON FIGURE 1.4.– The dragging technique originating in the Spanish Levantine coast, the *pêche au boeuf* spread from the end of the 17th Century over all of the north-occidental coasts of the Mediterranean. Quickly accused of destroying resources, *pêche au boeuf* was a reply to the increasing demands of urban markets. As they only required modest boats powered by the wind, the practice of *pêche au boeuf* progressively imposed itself over *pêche à la vache* and "tartanon fishing" (on the left of the etching).

Figure 1.5. *Fixed artisanal Mediterranean fisheries: the Tunisian charfia (source: Daniel Faget)*

COMMENTS ON FIGURE 1.5.– An ancient technique, *charfia* fishing relies on the use of traditional materials (palm trees and palm fiber fish traps) and the existence of strong community practices. The future of these artisanal fixed fisheries, which are nowadays threatened, illustrates the coexistence of traditional fishing and industrial fishing in the Mediterranean. The Tunisian *charfia* gives an example of sustainable fishing, with respect to the resource, orientated toward satisfying close alimentary markets.

1.1.4. *Speculative fishing: the madrague*

In use since Antiquity, the *madrague* is an immersed fixed post used to catch tuna. The fishing gear is presented as a large net divided into a succession of chambers, called "*le corps de la madrague*". The walls of this device are kept afloat by pieces of cork, their lower extremities reaching the bottom due to weights made of large stones (the Provençale *baudes*). This underwater trap, sometimes set at a depth of 45 m, is linked to the coast by a net that acts as a barrage: "the tail of the *madrague*", which guides the fish toward the entrance of the successive chambers. Set out in the months of February or March, for a fishing season that goes on to the start of autumn, the *madrague* is directed by a *rais* or *ray*, which organizes its exploitation. Highly developed in the modern era, the *brassier* are present all along the coast from Sicily to Gibraltar and from the Gulf of Gabes to Tangier. The Provençale coasts had no fewer than 19 of these fixed fisheries on the eve of the French Revolution. The big investments necessary for their exploitation (sailor's wages, net mending, conditioning and sale of fish, transport and *esparto*)[10] meant that they mostly evaded the fishing communities. They were headed by powerful landowners, such as the lords of Bandol or the Prince of Rohan in Provence [BUT 98] or the dukes of Medina Sidonia in Andalusia in the 18th Century [RAV 03]. These landowners, whose fishing rights came from royal privilege, did not look after their *madragues* themselves. They delegated their exploitation, by leases of three or six years, before a lawyer, to rich representatives of the world of negotiation and banking, who themselves worked for urban clients on behalf of their companies. In terms of price, the cost of a single *madrague* in the mid-18th Century was around 20,000 *livres-tournois*,

10 Originating in the arid regions of the occidental basin of the Mediterranean, *esparto* or *alfa*, also called *auffe* in Provence, was used to make ropes and baskets.

which is the price of a 40-ton tartanon [BUT 00]. The management of these fisheries is, therefore, beyond the communities, who do not consider the world of *madrague* to be true fishing. Calling the *madrague* societies "companies" and "firms", the members of the brotherhoods were well aware of what differentiated the world of the boat from these speculative and precapitalist activities.

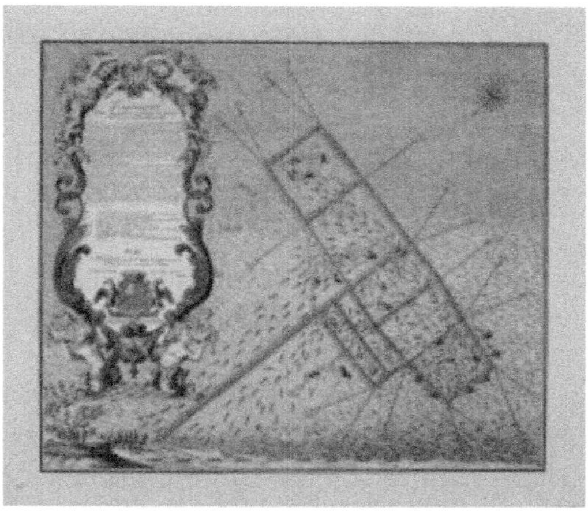

Figure 1.6. *Madrague etching (source: Archives départementales des Bouches-du-Rhône, 1Fi3398)*

COMMENTS ON FIGURE 1.6.– Very common in the Mediterranean in the 18th Century, the *madrague* were fixed fisheries aimed at catching tuna. Due to the amount of investment required, they were most often beyond the control of fishing communities, whose members had to work for them as employees. With high tonnages, they supplied a very prosperous activity of fish conditioning. Eight of these *madragues* are actually still in use in the Mediterranean.

1.1.5. *A highly varied consumption, reflective of relative shortages*

Coming from small-scale fishing, or from these large companies that are the fixed fisheries, the products offered at the fishmonger's stall in the modern era are extremely varied. Few species manage to evade capture according to the testimonies of naturalists of the 18th

Century. While compiling his inventory of Marseilles fishmongers in 1768, the German philosopher Martine Brunnich referred to no fewer than nine species of shark and catshark, including porbeagle, despite its reputation at the time as a man-eater [FAG 11]. Duhamel du Monceau revealed that in Provence "highly rated sausages" are made with the flesh of bottlenose dolphin, i.e. a large dolphin (*Tursiops truncatus*). In this diverse palate, the south coast of the Mediterranean is no exception. As in Sardinia, the consumption of sea turtles seems to be the norm on the Tunisian coasts, and we can find on these coasts, as on the coasts of Provence or Cataluña, the use of sea anemones (*Actinia viridis*) in cooking. However, all the available products are not equal. Michel Darluc, author of *Natural History of Provence* (published in 1786) made the distinction within marine animals between those species aimed at the people and those targeted at wealthy amateurs. On the fishmonger's stalls, the sardine, the bogue and the goldline porgy thus join the category of lower foods with the likes of the octopus, cuttlefish and squids, mollusks reduced by the author to the hardly appetizing rank of "cartilages", while more sophisticated buyers can share the more delectable sea dates and marinated tuna, with a special place for the whiting, king of the tables of Provence in the modern era. Behind this apparent health of the fish market in the 18th Century, there is the reality of relative shortages. The necessity of a certain level of protein in diets governs all here. Coastal populations also make the most of coastal fishing to complete their daily rations, by gathering limpets, periwinkle and green crabs – easy prey for the amateur fisherman. While most of the products of the sea used for food seem to be identical over the entire Mediterranean, detailed analysis shows that there are certain cultural differences in the art of using fish, mollusks and shells. Grilling is the most prevalent in the east of the Mediterranean, while boiling seems to be the rule in the septentrional coasts of the occidental basin.

1.2. Evolving practices (18–19th Centuries)

The complex world of Mediterranean fishing was not an unmoving universe over the centuries that preceded the birth of mechanized vessels. The product of migrations and an ever-increasing demand from the markets, the techniques spread over the entire basin, changing the faces and practices of fishing communities. While some of the techniques date back to the Middle Ages, their progressive use from the first decades of the 18th Century allowed a more intensive exploitation of the environment. Combined with an increase of marine pollution made worse by the first by-products of the industrial revolution, the acceleration of the exploitation of the resources was felt very early on. The fear of a depopulation of the seas and its corollary, the belief in a past golden age, appeared in intellectual writings from the end of the "Age of Enlightenment". These two themes became certainties in the 19th Century, when public powers and populations saw a confirmation in the rise of the price of fish and a rapid impoverishment of coastal marine fauna. The search for alternative solutions is explained by this context of worry. Mostly unnoticed by those involved in traditional fishing, aquaculture companies did flourish under the Second French Empire, but ended in failure, largely explained by the scientific limitations of those initiating it. As they multiplied at the end of the 19th Century, the research carried out in marine biology centers did, however, afford a better understanding of the dynamics of species, while the world of Mediterranean fishing was being transformed by the applications of mechanical power.

1.2.1. *The evolving world of fishing: human migration and technical changes*

Even more porous than national boundaries (themselves established rather later), the maritime zones are areas of active movement of men and knowledge in the modern era. Certain fishing communities played an important role in these transfers in the 18th Century. Among these vectors of technical evolutions, the fishermen of Provence or Cataluña appear to have played an important role in this period. Following the rapid spread of the sardinal in the 16th Century, the massive increase in dragged nets appeared, on the coasts of the Levantine seas, as the great event of the end of the 17th Century. This technique was essentially practiced from the coast up to this era. In Provence, it was the process of eissaugue, and in Languedoc it was the process of the *boulier*. The use of dragging nets in the sea from a vessel was still unknown at this time, since a royal decree from 1584 banned the use of the *dreige*[11] in the Atlantic. Regularly practiced in the 17th Century despite the ban, sea dragging was developed at the end of the century from Andalusia to the Adriatic, linked with the spread of the tartan, this new boat of roughly 50 tons, developed at the start of this century in the naval building yards of Martigues. This type of fishing, whose provisions were invaluable for the urban markets, was finally authorized under certain conditions by the decree on fishing of 1681. Its variants in Provence and Languedoc, *gangui, tartanon* or moulinet, presented a common aspect of being carried out by a single vessel, anchored (for the moulinet), drifting (for the gangue and tartanon). The spread of the *pêche au boeuf* (the Catalan *bou*), known from the Middle Ages in Valence, radically changed the characteristics of trawling. Now practiced by a couple of vessels, using the sail to tow a particularly large net, this technique considerably increased the efficiency of fishing. Only requiring small vessels, they democratized the access to the trawling art, freeing fishing owners from the obligation of possessing a costly tartan. It also allowed an extension of the fishing season; the *pêche au boeuf* was practiced with wind coming from the rear, as opposed to the *tartanon*, which placed the vessel across the wind.

As soon as it appeared in the Occidental ports of the Mediterranean, the *pêche au boeuf* was immediately adopted, resulting in a rapid expansion over less than 3 years from the north of the Valencian country to the Adriatic. Carried by Catalan or Provençale fishermen, then locally relayed in Italy by heads of the kingdom of Naples and Chiogga (Venice laguna), the diffusion of this technique illustrates the flexibility of Mediterranean fishing methods in the past. Present on the coasts of the Roussillon as soon as 1725, "pair trawling" reached the coasts of Languedoc in 1726. The following year it reached the Rhône, then appeared in Gênes and Livourne, where it developed in the direction of the coasts of Latium and Campania under the impulse of the fishers of Gaète, despite a ban issued by all the states concerned.

The initial pole of the diffusion of the *pêche au boeuf*, the Catalan coast provided an example of fishing communities powerfully

11 Dragged net.

transformed by new economical and social conditions which were those of this part of the Iberian peninsula from the end of the 17th Century. While its population doubled between 1713 and 1787 [HUR 02], Cataluña experienced a form of preindustrialization that led to an increase in the demand of food. The diffusion of *pêche au boeuf* can, therefore, be considered to be an answer to the growing needs of new urban centers in animal-based protein [ALE 03].

1.2.2. *Between the thought of decline and fragility of the environment*

As early as the end of the 18th Century, the rise of towing gears deeply destabilized coastal communities. In its very region of origin, the affirmation of the *pêche au boeuf* caused the emigration of part of the fishermen using fixed material, such as the Catalan longliners, who at the same time left the ports of Selva or Maresm to cast their lines in the waters of the gulf of Marseilles. The use of the *pêche au boeuf*, progressively adapted by the most enterprising of fishing owners, contributed to the division of communities already affected by the arrival of foreign bosses in their waters. In Marseilles, the divorce seemed confirmed by the end of the French Revolution between those practicing polyvalent fixed fishing and the fishermen converted to the practice of the "great art", who went as far as no longer recognizing the authority of the *prud'hommes* in the gulf [FAG 11]. This crisis in the communities favored the emergence of talk of decreasing populations, encouraged the recriminations of fishermen now marginalized by the trawlers. Scientific literature provides substance to this discourse, with the example of the work by the Dominican priest Antoine Menc on the origins of the decreasing populations of the Provençale waters [MEN 69]. This theme of decline was gladly taken up by the popular press and by some of the political elite during the 19th Century. It found a semblance of confirmations in the sudden increase of the price of fresh fish in urban markets, which progressively convinced the coastal populations of the reality of an impoverishment of the coastal ecosystems. It also explained the birth, in art and literature, of the figure of the poor fisherman, who is treated with the same compassion usually reserved for those on the brink of extinction. This idea of the end of a golden age is not new – the poems of Hesiod as early as the 8th Century BC, as well as the Old

Testament writs, have testified, since Antiquity, to the certainty of a cycle of decline in the West .

In the absence of reliable statistical series for the 19th Century, it is difficult to measure the reality of this impoverishment of the marine resources in the Mediterranean. The best that can be done is to note the disappearance of certain species from certain uses. In the Gulf of Marseilles, victim of excessive land-based fishing, the date mussels (*Lithophaga lithophaga*) disappeared was as early as the 1830s, at a time when most of the natural oyster fields were ruined on the Languedoc and Provençale coasts by the use of dredges [FAG 07]. Half a century later, the oceanologist A.F. Marion noted the progressive reduction of the size of the fish sent to Marseilles from the gulf of Algiers, a reduction that he attributed to the unchecked development of the dragging arts in the maritime space [FAG 11]. The techniques of fisheries in use adapted to these changes of the available resources. The use of the longline, therefore, became a rarity from 1850 in the gulf of Marseilles, while at the same time the municipalities of Sète or of Collioure denied the extreme rarity of the whiting in their fish auctions. The establishment of this very exact approach reflects the sense of urgency that there would be in the current research to orient the efforts of historians toward the establishment of databases on fisheries in the past. Still indicative (but are the tonnages carried out officially today by the fishing sector on a global level not themselves indicative?), these reference statistics could eventually avoid a number of biologists and fishery experts over using the *shifting base-line*[12], which does not allow a correct evaluation of the real evolution of the stocks available over the average length of time. The example of bluefin tuna, which has exceptional archives, such as the work carried out by the team led by Daniel Pauly, is a demonstration of the real possibilities that are available to biologists and sea historians, if the specialists allow themselves a certain level of interdisciplinarity [RAV 03].

As much as the words of political authorities, the lamentations of the fishermen also deserve to be examined. The recurrence, starting

12 Considering a decrease in population levels compared to a reference situation that is already damaged and not pristine.

halfway through the century, of complaints made by the *prud-hommes* against marine mammals, which are direct competitors of fishing, is indicative of the social and economic difficulties that affect the communities during this period.

Regularly mentioned in the texts of the modern period, the disturbances to fishing caused by marine mammals became a central theme in Mediterranean fisherman from the start of the 1850s. The affirmation of this near-obsessive hate provides an abundance of correspondence between bosses and representatives of the State or of local collectivities over most of the coasts of the Occidental basin. Claiming that the multiplication of small cetaceans was ruining fishing, the fishermen received concrete measures from the public powers. These were added to the rewards offered by the *prud'hommes* to fishermen able to prove the destruction of a porpoise or a dolphin. With the help of the Navy, anti-cetacean systems were also trialed after 1890 over the entire Provençale and Languedoc coast, whether with trapping baits, exploding nets or cannons, whose effectiveness is very hypothetical. This hostility toward marine mammals did not stop with the start of the 20th Century, since in 1920 it was even reflected by aerial bombings carried out on the Algerian coast. In Sète, if we do not consider the individuals killed by bullets or cannons, nor those victims of the deadly baits, 254 small cetaceans were eliminated between 1904 and 1910. Moreover, the port of Sète is not the one in which the elimination campaigns were carried out with the most vigor. Thousands of dolphins were undoubtedly killed in the Mediterranean coasts before World War I. These massacres are surprising for the contemporary readers, for whom the new sanctification of the dolphin is a familiar reality; they provide us with a double lesson on the fishing communities of the 19th Century. On the one hand, it draws our attention to the increased sense of vulnerability felt by the fishermen during the century of industrialization. This can be explained first by changes in the methods of fishing. The use of the roe by sardine fishers, extremely rare in the 18th Century, is noted by certain *prud'homies*, such as Martigues from the 1830s onward. This usage certainly increased attacks of dolphins against quickly damaged fishing material. The increasing competition from Atlantic fishing, which arrived massively on the Mediterranean coast with the opening of the Paris-Lyon-Mediterranean line after 1850, only increased the worry of communities that felt left aside by the social changes caused by industrialization. The reduction of the acceptability of the constraints caused by marine mammals finally marked the birth of a new way of looking at the animal world, with the distinction between those that are useful and those that are a nuisance. It is through this that the maritime authorities and the oceanologists at the beginning of the 20th Century sought to establish an industrial branch of dolphin fishing which, by commercializing oils and fats, would make the massacres profitable. Taking inspiration from porpoise fishing in the Black Sea, very active throughout the 20th Century, the idea developed by the Marseilles biologist Paul Gourret fitted into this process. The mastering of processes of refining petrol, by allowing the production of low-cost mineral oils, would, however, make these projects of industrial fishing obsolete after World War I.

In this context of worry and increased competition in the exploitation of the maritime medium, Mediterranean fishermen also had to confront the first effects of industrial and urban pollution. It is often wrongly assumed that maritime pollution in the Mediterranean started in the first half of the 20th Century, when the industrial release of chemical molecules started to become massive. However, locally, we can observe real changes in the ecosystems of the infracoastal layer caused by factory waste as early as the end of the 18th Century. In Marseilles, over 12,000 tons in total of polluting residue were released annually into the seas in rocky inlets close to the port by soap production from the 1780s. Their effect on the maritime environment was catastrophic, as shown by the inquests carried out by officers of maritime affairs from 1820. The development of the chemical industry and non-iron metal industries worsened the situation in the gulf throughout the 19th Century. The case of the factory belonging to the exploitation Company of Rio Tinto, established on the Estaque coast in 1882, is emblematic of this, since it is reflected in the near total ruin of professional fishing in the north of the commercial port of Marseilles. On top of this chemical pollution can be added millions of tons of organic waste that dangerously damaged the quality of the waters of the biggest ports of the Mediterranean, such as Marseilles,

Geneva, Barcelona or Algiers, to the point of making the consumption of shellfish caught close to certain urban centers dangerous [FAG 05].

Faced with the difficulties of coastal fishing deemed incapable of properly handling the resource, entrepreneurs and scientific institutions tried throughout the 19th Century to restore the medium's ability to deal with the food requirements of coastal populations.

1.2.3. *Aquaculture, or the dream of a domesticated sea (1850–1900)*

From their origins, aquacultural companies escaped the control of *prud'homies* and brotherhoods. They appeared as an application to the maritime space of the desire of free enterprise developed by liberal thinking. Its protagonists, first of all motivated by the hope of solid profit, provided an activity that was of interest to the public. For this reason, they therefore benefited from constant support from the State, with the Second Empire of France representing a real golden age for aquaculture experimentation. Coming from various horizons, since this group was made up of "sea industrialists", as they liked to call themselves, real scientists, such as the professor of compared embryogeny at the College de France, Victor Coste, powerful industrialists, Léon Vidal, for example, or true adventurers, like the representative of the Imperial acclimation Society Eugène Lamiral, the aquaculturers expressed a common faith in the thoughts of Saint Simon. Without doubting man's capability to cultivate the sea, they sought to transform traditional fishermen into gardeners of the marine space, for the greater good of the consumers. Their initiatives, of various forms, are a testimonial to the fascination felt by these men for the shell farms of Lake Fusaro and of the little sea of Tarente in Italy, where most of them traveled as part of their studies during the 1850s. Designed as an answer to the ruin of the oyster benches in the Languedoc, the trials in oyster cultivation and reproduction conducted by Victor Coste from 1860 onward in the Thau laguna had massive potential in the eyes of these entrepreneurs. Inspired by this, Léon Vidal started experimenting with the culture of mullet and sea bass from 1864. He conducted trials of castration or artificial insemination, without hesitating to call his activities "stalls or courtyard cultivation". Although the trials conducted by Léon Vidal, like those of Victor

Coste, ended in failure, largely attributable to gaps in scientific knowledge at that time, they are of much interest for the history of fishing. By proposing to employ the fishermen of the Martigues as simple providers of feed to be used to fatten the fish in his farms, Léon Vidal was designing the outline of a new organization of fishing, marked by an increased dependence of the communities faced with private interests external to the world of traditional fishing. By planning to sacrifice part of the marine resources taken from the Berre laguna to his cultures, Léon Vidal announced the development of the current milling fleets, entirely dedicated to the service of aquaculture farming. The example of Coste and Vidal finally planted the seeds of future developments in Mediterranean aquaculture, which would finally be successful in mastering the cycle and production of oysters in the Thau lagoon at the beginning of the 20th Century [GIO 95].

1.2.4. *Rationalizing the fishing industry in the Mediterranean: the rise of "applied zoology" (1880–1914)*

From the beginning of the 1880s, coastal Mediterranean fishing became the object of the studies of oceanologists. From the Balearic Isles to Sicily, laboratories of marine biology looked into the best way to help communities to better manage the resource. In France, the rise of the station of Banyuls-sur-Mer, Marseilles or Villefranche-sur-Mer, to only mention the most important ones, happened in parallel to the reconstitution in 1887 of the technical service for maritime fishing, which was first of all an answer of the public powers to the worry expressed in several French ports as a result of a reduction in the tonnages carried out. Created by the decree of 17th May 1887, this service was made up, as of 1862, of a general inspection and an advisory council, forming the advisory committee of maritime fishermen, answering to the minister of the Marine and the colonies. From its creation, this organism was perceived by some of the oceanologists as an efficient one, likely to provide a method of immediate action to reform fishing in the Mediterranean. The Endoume laboratory in Marseilles appeared in the sense as the main place for reflection on this subject, under the watch of Professors Antoine-Fortuné Marion and Paul Gourret. By starting in 1888 works of "applied zoology", these two men began a phase of fruitful cooperation between fishermen, public powers and scientific environments. This can be essentially separated into three axes [FAG 11]. Partisans of a reconstitution of the resource by preservation of part of the maritime, the Marseilles scientists obtained in 1864, with support from the *prud'homie* of fishermen, the creation of a cantonment zone, which we can consider to be the ancestor of the marine parks in the Mediterranean. A site of observation favored by oceanologists, the cantonment area of Endoume provides the proof of the capabilities of rapid reconstitution of the resource in the absence of fishing.

Open to the necessity of high-level training for fishermen, Paul Gourret proposed furthermore in 1893 the creation of an institution for the professional training of future fishermen. The world of Provençale fishing therefore anticipated, quite early on, the foundation in February 1895 of the *Société d'enseignement professionnel et technique des pêches maritimes* (SEPTEM) (Society for the Professional and Technical Training of Maritime Fishing), whose first director was Emile Cacheux. Officially opened in 1896, the professional school of maritime fishing of Marseilles provided for its students (aged 11–20) technical knowledge, the handling of materials, elementary principle of navigation, as well as basic knowledge concerning marine ecosystems and a better understanding of their vulnerability, which was completely new. Going further than just the Marseille region, Paul Gourret compiled a manual in 1898 for use by primary school students along the French coast, called "The sailor and the fisherman. Lessons relevant to the marine profession and to the fisherman", which showed his desire to improve the practices carried out by the fishing industry through education.

Eager to associate the population, which they considered to be a vital partner for the proper management of maritime media, Marion and Gourret finally opened a fishing museum in 1897, in which not only material and fishing vessels were exhibited, but important displays dedicated to living species were also presented, as were their uses by human societies. Even more innovative, as it is linked to the first steps of ethnological science, the fishing museum of Marseilles finally presented elements that constituted the traditional culture of the fishermen of the Mediterranean, idioms, beliefs and traditions.

If the example from Marseille appears representative of this new path of Mediterranean fishing, now followed and advised by scientists, it is not alone in the entire basin. At the same time, the works of Raphaël Dubois on sponge fishing and Mediterranean pearl oyster fishing in Tunisia, of Armand Sabatier on oyster farming in Thau laguna or those by Odón de Buen on the management of the management of fishing resources on the Spanish Levantine coast showed that the world of the boat was entering a new age on the eve of World War I. The mechanization of vessels, the rise of deep fishing, the dilation of active zones to the colonial domain and the revolution of refrigerated transportation at this date opened up new horizons for most of the fishermen.

1.3. Industrial power at the service of fisheries (end of 19th–20th Century)

The industrial revolution that animated Europe and North America from the end of the 19th Century triggered a chain reaction that dragged along all the sectors of economic activity.

First, the use of steam, then of new energies, such as electricity[13] and petrol, would stimulate the development of the textile industry and the steel and metal industries, with notably the mechanization of the methods of production and motorization of conveying. Steam-based navigation and railroads would close the continents, reduce the distance from the coast and the duration of travel. The development of methods of communication with neighboring cities would favor the development of seaside tourism and coastal economic activity [AMO 60]. The application of the work of Charles Tellier on the conservation by artificial cooling (1867) would ensure the transport of perishable goods to further away places, thus contributing, with the

13 "[…] steam, electricity will have allowed human ingenuity to make, in less than half a century, a huge step in the direction of progress, that up to then, would only have been followed step by step" [LED 35].

improvement of methods of transport, to the increase in the amount of consumables available.

This demand for foodstuffs, and notably marine products, becoming increasingly pressing under the effect of expanding demographics and the improvement of feeding habits, would incite politicians to support any innovation in favor of the development of the most productive segments of the fishing sector.

These innovations toward greater productivity always met resistance from the traditional knowledge of the fishermen, forged through often empirical experience of the sea built over several generations. The tradition opposed itself to any risk of the change to the social and economical balance of their community, favoring the knowledge of the "man of art" over that of the "engineer". Up to then, the transmission of knowledge was essentially the fruit of the migration of the fishing communities around the Mediterranean basin [BAS 05, MER 63].

Most of the resistance, however, would cease faced with unequal competition with more efficient methods and the hope of greater productivity promised by the perspective of exploring greater areas of sea [FER 01].

The transfer of these innovations to fishing would be made unequally. Alone, regions that already benefited from a level of industry, established communication networks and a workforce were the first to benefit from this. The fisheries of maritime regions of countries with no industry would encounter a lesser and later economic growth, essentially resting upon the organization of the traditional societies of fishermen (*prud'homies* and brotherhoods).

With the industrial revolution, two types of organization of fishing determined themselves, one being not very territorial and more orientated toward maximizing fishing quantities, the other, established on traditional knowledge and developing a more patrimonial economy [CEN 01, FER 01].

1.3.1. *The industrialization of Mediterranean fishing*

In the Mediterranean, industrializing would mainly affect mass-production fishing, aimed at preservation and export, i.e. first *madragues*, and then other fishing of pelagic species and finally trawler fishing. However, with the exception of some shipping companies operating in the Atlantic during the 1950s and the 1980s [BAR 61, FOL 93], this industrialization is not total and justifies the denomination of semi-industrial fisheries [LLE 98]. While the copropriety between several fishermen of production methods, or the financial participation of fish traders, does exist, Mediterranean fishing is in most cases carried out within artisanal- or familial-type structures, the industrial character being based mainly on the marketing of the productions.

The other jobs commonly called the *petits métiers*, or also incorrectly "artisanal fishing", are very poorly taken into account by policies of development. Their modes of production are more orientated toward self-consumption, exchange and direct local sales, and, through modest profits and high diversity, correspond badly to the criteria of quantity, monospecificity and regularity of the industrial markets. Until the 1950s, the simple and rudimentary character of their catching techniques meant that their fisheries benefited very little from the technological progress brought by industrial development. It was only later in the second half of the 20th Century that nylon, outboard engines, composite flat hulls and geo-positioning system (satellite-based GPS) ended up giving an extra boost to these small fishing jobs, by allowing them to move away from the coast to look for higher value fish [BOU 87].

The technological progress that benefited semi-industrial fishing touched all of the functional components of the fishing industry, from the fish production to cannery, passing by the textile industry and naval construction. Unlike agriculture perhaps, few innovations were specifically dedicated from their origin to fishing. Most of them came from the adaptation of systems, tools and materials created for other sectors of the economy, such as marine or train transport, or the industries of weapons, clothing or food. Applied to fishing, they have provided constant improvement to four main domains: energy and vessels, communication and information, fishing gear and its handling, and the conservation and the valorization of the catch. These technological changes would follow different paths over time depending on the fishing methods and the evolution of economic strategies. The evolution of Mediterranean pelagic purse seining, as well as the evolution of trawling, illustrated this over the last two centuries.

Figure 1.7. *Catalan Lamparos (source: Jacques Sacchi, 1973)*

COMMENTS ON FIGURE 1.7.– The technique of fishing sardines with light is one of the most widespread in the Mediterranean since Antiquity. The word "lamparo" refers to both the technique and the lighting devices, the vessel and the encircling net used to catch the fish. By replacing the sardinal fishing technique, this method of fishing strongly contributed to the sardine industry of the Mediterranean.

1.3.2. *Seine fishing*

Whether bluefin tuna (*Thunnus thynnus*) or sardines (*Sardina pilchardus*), investors and politicians quickly saw the profit they could draw from these species, excellent products of commercial exchange through their abundance and nutritional value. The birth and the history of the preservation industry in the Mediterranean are strongly linked to those of the fishing of these species. Previously preserved and sent off in a wooden barrel, sardines and tuna would be the first to benefit from the invention by Appert [APP 13] and from the use of tin cans ("tinplate") to the benefit, first of all, of armies and the French navy. In its beginning, this industry that only required a large, non-qualified workforce required no particular technology at the beginning of the 20th Century, except for the soldering and setting of the boxes. The rapid success of this industry quickly made the vessels dependent on the preservation market, which contributed to the variations in the fishing landings. From its industrialization onward, pelagic fishing in the Mediterranean went through a series of crises linked to irregularities in stock availability, exploitation costs, drops in consumption or to competition with Atlantic products [BER 87, MET 63, RIO 05]. During each period of recession, the companies and the public converted to the new technology available, without being able to resolve the chronic instability present in this sector.

Drifting nets (i.e. "*sardinal*", "*courantille*" and "*thonaille*") have progressively disappeared and been replaced by encircling techniques (thys and traina) [GOU 97] and turning seines (*lampara* and *lamparo*). The *lampara*, appearing at the end of the 19th Century in Italy, most likely originated from the ancient big coastal seines, as noted by their "spoon" shape, the central pocket and the wings that characterize them. "Lampara" fishing consists of encircling at night the shoals of blue fish grouped together by the light, and to manually hauling the hemp or cotton net, of 150–300 m in length. Much more efficient than the drifting nets, the "lampara" developed around the Mediterranean due to migrations of Italian fishermen, leading to an important increase in production and opposition and worry from traditional fishermen, fearing that more productive fishing would cause a sudden decrease in the value of their produce [BAR 61, SIM 61]. Sardine fisheries are often confronted with a drop in their values and a decrease in catches. Under the threat of the closing of the canneries and dramatic social crises [MAR 97], these can only be compensated for by imports of fish or by an increase in the fishing effort. After

World War I, and under pressure from factory workers, the public powers then favored the introduction of the *ring net*. This evolution from the Italian "lampara", imported from California at the same period, presented on its lower part a series of rings in which ran a closing cable. First introduced in Italy (1925), and then in Spain (1936) [BER 87], the ring net became, by the end of the 1950s, after several modifications including the shifting of the bag to one of the extremities and the use of nylon instead of cotton, the main piece of fishing gear for fisheries of small pelagic species in the occidental Mediterranean [DIE 53]. Due to the introduction of the *power-block* "puretic", a sort of mechanized wheel for the raising of the net, the mechanical capstan and the generalization of the use of ultrasonic sounders, sardine purseiners could now haul in 9 min with five men a net of 200 m that required over 10 men with the *lamparo* [MAU 61].

Tuna fishing was also revolutionized by the arrival of this new method of capture. The *madragues*, remaining for centuries the main method of capturing red tuna, progressively gave way to mobile techniques such as the "seinche", a collective technique that remained in Provence and in the Languedoc until the arrival of the turning and sliding tuna net at the beginning of the 1960s [RAV 11]. The development of the purse seine in the Gulf of Lions is largely linked to the arrival of displaced fishermen from North Africa using their knowledge of the *ring net* and its variants for small pelagic species to adapt it to tuna fishing.

The "catalanes" vessels and the "pointus" were, little-by-little, replaced by bigger units of 18–20 m, capable of using the turning sardine net or the tuna "seinchole" depending on the circumstances. From the beginning of the 1970s, due to the difficulties felt by the sardine market, part of the French purse seine fleet became specialized in the fishing of bluefin tuna, which was more promising. Under the impulse of plans of restarting and modernizing fishing started by the initiative of regions and governments, and then by the European Union (EU) beginning in the 1980s, construction was developed and became standardized. The vessels adopted in 1973 the "clipper" shape, with a wheelhouse at the front and a *flying-bridge* to ease detection. They were extended, reaching roughly 4 m in 1990. With the customizing of hulls for the preservation of catches, their capacity increased sixfold in 40 years. The replacement of wood by composite materials in 1975 lightened the hulls, while greatly decreasing maintenance costs. With propulsions three times greater than those of the 1960s, they reached a maximum speed of more than 17 knots in the 1990s. Starting from then, their design started to largely resemble those of ocean tuna vessels, with the first constructions carried out in 1990 in the Atlantic shipyards and the installation in 1997 of a ramp on the stern for hauling the skiff (working boat). The equipment of the bridge also increased ergonomically, the capstan of the 1960s being replaced with a purse line winch due to changes in hydraulics, capable of closing the purse seine in less than 15 min.

In the meantime, the conception of tuna purse seines continued to progress notably due to improvement in hydrodynamic calculation in the 1990s. At 600 m in length in the 1960s, tuna purse seines reached 2,000 m 40 years later, increasing their deployed surface eightfold.

It is in the domain of communication and detection that pelagic fisheries obtained the most benefits. Initially used as a link with the mainland or for security reasons, the information and communication technologies were adapted to various functional ends by the Mediterranean fishing industry at the beginning of the 1960s.

As for all pelagic fishing, bluefin tuna fishing required long hours of often unfruitful prospecting (70% of the fishing time for purse seiners), which naturally involved cooperative work. Communication between associated vessels and the protection of the transmitted information from competing vessels appeared, therefore, from the start of this technique as a vital factor for success. This legitimate aspect of the use of single-sideband modulation (SSB) and very high frequency (VHF)[14] radios from the 1960s onward was followed by the satellite facsimile at the end of the 1990s.

14 Single-sideband modulation (SSB) is a type of radio transmission used on boats for long-distance communication, unlike very high frequency (VHF) whose reach is limited to about 30 nautical miles.

Bluefin tuna being first of all a surface fish, its detection by sight is primordial and any technology will look to give the greatest field of vision. Thus came aerial prospecting in 1975, which goes from the simple detective aerial flyover to a way of helping the purse seine shot that is so efficient that it ended up being banned by the International Commission for the Conservation of Atlantic Tuna (ICCAT) in 2006. The use of highly discriminating radar to detect the flights of birds flying over tuna shoals (S-band radar or "bird radar") and the extension of the use of sonar from the 1990s onward provided the vessels with a considerable amount of effectiveness in the detection of fish shoals. The democratization of GPS in 1994 finally ensured greater precision in fishing operations and navigation. All the while improving the effectiveness of the fishing and commercialization operations, this technology has, therefore, contributed to producing a greater fishing effort.

Up until the mid-1970s, bluefin tuna was destined mainly for the Mediterranean fresh fish or cannery markets; its storing onboard required only icing or brine for long trips. From the mid-1980s, due to the Japanese "sashimi" market, this fish acquired a high market value, requiring more sophisticated preservation methods, such as immersion in refrigerated water (1994) and deep freezing, and larger vessels as a result .

The exponential size of the demand quickly resulted, from the 1990s, in a radical transformation of the system of catching and preservation. The near-totality of catches are thus no longer brought aboard, but rather transferred directly to floating catches that are dragged to coastal fattening zones. The extension of this so-called "opening up" technique to the whole of the Mediterranean led to an excessive development of fleets such that, out of fear of a collapse of this resource, the ICCAT and the EU imposed reduction in the quotas for catches in 1998 and reduction in the sizes of fleets [CIC 06]. These increasingly drastic measures have led to a subsidized removal of the least profitable tuna seine boats, too specialized to be reconverted into another job.

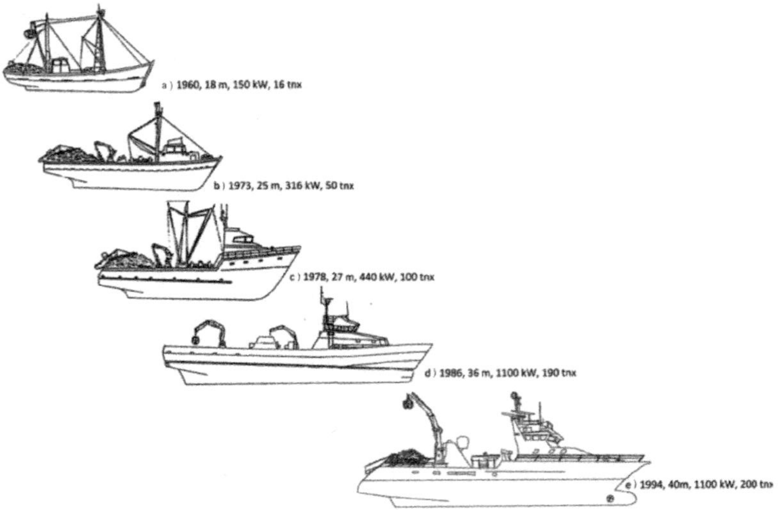

Figure 1.8. *Evolution of French tuna purse seiners in the Mediterranean (source: a)–c) Henri Farrugio and d)–e) Jacques Sacchi)*

COMMENTS ON FIGURE 1.8.– Since the 1960s, the length, tonnage and engine power of tuna purse seiners have not stopped increasing, contributing to an overall increase in the fishing capacity of the fleet, beyond the limits of sustainable exploitation of the bluefin tuna resource.

Figure 1.9. *Modernization of vessels and equipment (source: Jacques Sacchi, 1973)*

COMMENTS ON FIGURE 1.9.– The specialization that took place from the 1960s in the practice of "métiers" calls for a more ergonomic reorganizing of the fishing bridges, mechanization, the use of hydraulics for hauling gears (winches and *power-block*) and electronics in navigation and the spotting of shoals. This modernization contributed toward the improvement of the conditions of navigation and work at sea, as well as greatly increasing the effectiveness of the fishing power of the vessels.

Figure 1.10. *Bluefin tuna purse seining in the Mediterranean. Transfer of a catch to towed floating cage (source: Henri Farrugio, 2008)*

COMMENTS ON FIGURE 1.10.– The tuna captured by the tuna purse seiner are transferred alive directly from the purse seine to a floating cage, which is then towed to the coastal fattening zones for the Asian raw tuna market; this technique, imported from Australia, has reanimated the Mediterranean bluefin tuna market by optimizing the effectiveness of this branch and the quality of the catches.

1.3.3. *Trawl fishing*

Over the last two decades, the development of the Mediterranean trawling fishing was essentially conducted as a unique response to the ever-increasing demand from neighboring countries for fresh and refrigerated products. With a few rare exceptions, these fisheries were concerned very little by the industrial valorization of their productions in the form of transformed products. They mainly focused on extending their field of action further and deeper by gradually improving their capturing efficiency and the quality of these catches. In a constant quest for unexploited, far away resources, the evolution of propulsion has been a determining factor in trawling and the vessel's ability to pull its fishing gear. Steam, therefore, appeared from the end of the 19th Century as the opportune answer to this demand for most of the fisheries of the North Atlantic. The Mediterranean was, however, less attracted by this mode of propulsion, the amount of coal necessary to fuel the machines weighing the vessels down considerably. Until 1920, the Mediterranean trawler fleets were, therefore, mainly composed of sailboats working in pairs [GOU 91]; only ports benefiting from sufficient infrastructure (Marseilles, Algiers, Bougie and Ancôna) could receive steam trawlers. After World War I, the Mediterranean trawler fleets quickly adopted the internal combustion engine[15] [BAS 55]. This technology appeared to be more adapted to coastal fishing carried out with vessels of low-to-medium tonnage. The serial release of the "diesel" engine (Ets Baudouin) with integrated reduction gearbox in 1929 was a decisive step. With an energetic yield far greater than the steam engine, the use of diesel, therefore, left more hull space for the cargo, by carrying less fuel. Due to the reducer, it also allowed for the use of less cumbersome, lighter and more powerful propulsive groups. The four-stroke engine, with turbo compressor, the variable step propeller (1994), the nozzle around the propeller would thus lead to an increase in the effective power of trawlers by 5–8 from the 1950s to the present day [EVO 11].

The development of the use of the diesel motor favored the development of the mechanization of winches. From the 1950s onward, the use of hydraulics brought more flexibility and security to maneuvers as well as greater traction power. With the stern trawling which started at the beginning of the 1960s and the replacement of the

15 By directly transforming inside the engine the thermal energy produced by a fuel into mechanical energy, internal combustion motoring allowed for the removal inside trawlers of the cumbersome boilers that were needed in steam propulsion.

cargo boom with a stern gantry, the hydraulically controlled net reel drum of the trawler, introduced at the beginning of the 1970s, marked a cornerstone for the trawler, by reducing the danger and difficulty of the maneuvers.

The hulls of Mediterranean trawlers evolved more slowly. At the beginning of the 20th Century, the large majority were made of wood in artisan yards. The creation of "deep water" ports from 1918 [MUS 21] allowed the receiving of vessels with higher tonnages. The hulls then took on a "V" shape, without a lateral keel, the vessels no longer needing to be brought to ground. As for pelagic fisheries, trawling in the Gulf of Lions greatly benefited from the repatriation of large units of North Africa at the beginning of the 1960s. It also benefited from regional, national and European plans for the modernization of fisheries. The vessels benefited, as did tuna purse seiners, from technological advances in the use of composite material and aluminium for their construction.

After navigation by estimation and manual probing, the functions of detection and navigation called upon increasingly sophisticated methods, in hydroacoustics with paper band probes in 1970 and with video in 1993, and geographical positioning with the Decca Navigation System in 1965 and GPS at the end of the 1980s. The miniaturization of the electronic equipment and information technology resulted in, from the second half of the 1980s, the use of electronic navigation charts. Coupled with a satellite positioning system and numerical bathymetrics, these charts contributed to creating increasingly precise fishing paths, with a better consideration of the substrate.

Being strongly linked to the pulling capability of the vessel, fishing gear has followed changes in the methods of propulsion. Originally imported mainly from Italy and Spain [GOU 91], the Mediterranean trawls did not see any fundamental modifications until the introduction of divergent panels toward 1898, which increased the horizontal opening, and the introduction of the Vigneron-Dahl trawl in 1928. These innovations thus gave a determining advantage to the single-vessel trawl. After World War II, the hemp ropes used in the construction of trawls were replaced at the end of the 1950s by thinner and more resistant nylon string. The design of French and Italian trawls benefited from this period from studies carried out in the North Sea on models in trial waters. These allowed improvements in hydrodynamics and behavior during fishing. Their vertical opening, mostly smaller than 2 m in 1973, went beyond 10 m in the 1980s, with fishing surfaces almost five times greater.

The introduction, from this period, of seabed trawlers with "large vertical openings" (LVOs) in the Gulf of Lions constituted a response to the stagnation of the production of seabed fish. It provided the market with a range of open water species, such as hake and small pelagic species that the traditional trawlers would rarely catch. In this spirit, to patch the irregularity of the supply of small pelagic fish by the purse seine, some Mediterranean countries, such as Italy [FER 81], Greece and then France [MEU 87], introduced trawlers with very large vertical openings (20–25 m), inspired by models created in Denmark in 1965 for herring fishing. In tandem, or with four diverging panels (Gulf of Lions), this new technique of catching sardines or anchovies experienced a rapid expansion due to the support of electronic prospection equipment (sonar and probes) and electronic trawler positioning systems ("Netzsonde"). It notably allowed for an improvement of the commercialization of the sardine and anchovy by spreading out the produce over more regular supplies [MEU 87], something that the Mediterranean technique of purse seining did not allow. However, it led to the construction of new, more powerful trawlers, but ones more economically dependent on the resource and market.

After more than 20 years of relative stability, the sardine industry has found itself again, from 2007, in a state of crisis. It must deal with a substantial shortage of resources and a rise in the price of diesel, leading to a relocation of the fishing effort of the trawlers to demersal species. This new crisis, which affects most Mediterranean pelagic fisheries, once again asks questions over the exploitation of unstable resources and the consideration of the influence of the variations of the surrounding environment on their abundance and presence (see *The Land–Sea Interactions [MON 14a]*, and, *Ecosystem Sustainability and Global Change [MON 14b]*, in the "Seas and Oceans" Set.

Faced with the reduction of yields on the continental plateau, the deeper resources of the continental slope, notably of red shrimp, have also represented a hope of redeploying trawler fishing [SIM 61]. This orientation does, however, require the use of specific equipment adapted to deep water fishing such as high capacity winches, deep water echo-sounders and an as precise as possible position of the fishing hauls. While as early as the 1960s, some Mediterranean trawlers were able to equip themselves with fishing winches equipped with 3,000 m of dragging cable, it was only in the 1980s that part of the Mediterranean trawler fleet started to orientate itself more openly toward deep fishing. Some of these units became specialized in the fishing of Norway lobster and red shrimp, on beds of 400–800 m, while others carried out a double activity of seabed fishing on the continental shelf and deep crustacean fishing, depending on the season.

Figure 1.11. *Evolution of the Mediterranean trawler fleets (Source: J. Sacchi)*

COMMENTS ON FIGURE 1.11.– After World War II, Mediterranean trawlers quickly abandoned the sail (a) for steam (b) and diesel (c), thus allowing them to extend their field of activity far beyond the 12 nautical miles exploited by the "*chaluts boeufs*" of the 19th Century. This evolution became accentuated around the 1980s, notably by replacing wood with synthetic materials and steel (d) for the construction of the hulls.

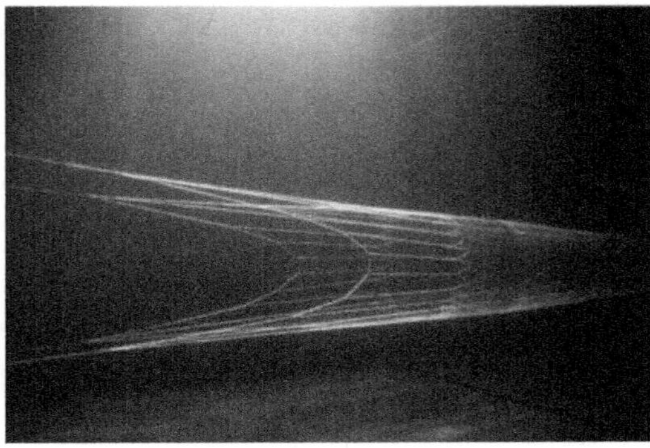

Figure 1.12. *Pelagic rope trawl, or "parachute trawl" (source: J. Sacchi)*

COMMENTS ON FIGURE 1.12.– This trawl for which the front and the wings are mainly made up of rayed ropes is one of the most recent creations introduced for fishing of small pelagic species in the Gulf of Lions. Its very large vertical and horizontal openings allow it to sweep over a section of water of more than 1,000 m^2.

1.4. Fishermen today in the Mediterranean

In 2008, without counting tens of thousands of small non-motorized boats, the number of declared fishing vessels active between the strait of Gibraltar and the Dardanelles has been estimated at around 82,000, of which 83% are small units of less than 12 m, essentially practicing a polyvalent activity of the small artisanal fishing. The vessels greater than 12 m are usually specialized in a limited number of techniques such as the towing (trawls and dredges), encircling (purse seines) and static (longlines, gillnets and fish traps) arts. The trawlers and dredgers represented around 10% of the Mediterranean fishing fleet, the purse seiners 4% and the longliners, gillnetters and potters 2%. To these fleets of motorized vessels can be added roughly eight *madragues* for the fishing of bluefin tuna. All of this fleet employed roughly 250,000 sailors, 55% of whom worked on vessels of low-scale fisheries. The engine power of all of the vessels was in the order of 6 million kW, 60% of which went to the trawlers, the purse seiners and the boats of static fishing of more than 12 m.

However, the countries of the EU had 58% of the Mediterranean trawlers and 26% of the purse seiners [SAC 11].

Fishing in the industrialized countries of the North Mediterranean has experienced a sustained decrease in its number of vessels and sailors from the first half of the 20th Century. Helped by the motorization and the mechanization of the vessels, it continued after World War II with a progressive decline of small-scale artisanal fisheries and, from 1983 onward, the implementation of European policies on the reduction of fishing abilities. On the contrary, the number of fishing units of the countries of the East of the Mediterranean and Northern Africa is slowly progressing, following in this sense the developmental policies of the countries involved [SAC 11].

Figure 1.13. *Small-scale artisanal fishers in Sète (source: J. Sacchi)*

COMMENTS ON FIGURE 1.13.– The vessels for small-scale fisheries represent more than 85% of Mediterranean fishing boats [SAC 11]; their activity is polyvalent and established for the production of varied species with the main target being regional fresh product consumption. Their evolution has benefited very little from technological innovation and has experienced, like most of Mediterranean fishing, a reduction in numbers.

The apparent stability of these means hides the reality of an increase in the effective capabilities of the Mediterranean fleets as it does not take into account the constant progress made in terms of the mechanization of fishing equipment, communications and detection; without forgetting the transmission of knowledge, hygiene and the improvement of living conditions at sea. The speed of the increase in fishing power has obviously been different between the north and the south or the east and the west of the Mediterranean, depending on the investment capacities of the fleets in technological innovation [SAC 11]. The first beneficiaries of progress, the same industrial fleets of the EU countries, therefore, saw their technical efficiency increase by 3% per year despite a decrease in their total number of vessels [DES 09].

1.4.1. *The role of fishing in the consumption of sea products*

Mediterranean fishing is mainly multispecific and polyvalent. While the Mediterranean trawlers do target mainly bottom fish with the gillnetters, crustaceans and mollusks of the continental shelf and slope, the purse seiners boats are, with some drifting longliners and the *madragues*, mainly orientated toward the capture of pelagic species, such as the sardine and the bluefin tuna.

The production of consumable sea products from Mediterranean fishing is estimated at slightly less than a million tons, according to the Food and Agriculture Organization (FAO), or less than 1% of global fishing. Established from quantities on the national markets, this estimation considers less than perfectly the produce from small-scale fisheries, mainly sold directly to local consumers.

Split into near equal proportions between demersal species and pelagic species, this production experienced a global stagnation of catches [GAR 11] from the start of the 1980s. Indeed, the production of the countries of the South, regularly growing since the beginning of the 1950s, caught up with the declining production of the countries of the North of the Mediterranean basin in the 1990s.

Sea fishing is only source of marine products for Mediterranean countries. There is also the increasing production of marine and lagoon cultures. A result of traditional shellfish farming and fish farming of freshwater and brackish water, aquaculture experienced a rapid growth from the 1990s onward, currently reaching more than a million tons due to the development of marine fish farming in the countries of the EU.

The Mediterranean market for Mediterranean fish was dependent on imports coming from Atlantic and North Sea fisheries well before the beginning of the industrial revolution. Relayed in the 20th Century by the diet culture and the development of maritime tourism, the popularity of low-fat foods born in the previous century led to a spectacular increase of the dependence of diets of the countries of the North of the Mediterranean on sea products [ETI 10]. In the countries of the South Mediterranean, the increase in the consumption of marine products remains small, however, largely dependent on the ability to ensure proper nutrition and on local customs.

Without direct data on the consumption of marine products in the Mediterranean, we do not have any other information apart from the evolution of the types of food available published by the FAO to demonstrate the evidence of this situation. This estimated consumption has gone from around 7 kg/person in 1961 to 14 kg/person in 2008. However, the portion of the production that represented around 40% of this consumption in the 1960s has progressively dropped from the beginning of the 1990s, to only cover 21%. The slightly larger portion of demersal fish in the consumption of marine species in the Mediterranean (61% in 2007) is mainly linked to growing interest in fresh fish and a certain amount of distrust for small pelagic species (anchovies, sardines and sardinellas), too quickly altered and instead aimed at preservation. This fascination with bottom species is one of the causes of the deficit in the commercial balances of Mediterranean countries; these only produce, taking into consideration aquaculture and continental fishing, about 70% of what they consume.

Generally, the preference of Mediterranean consumers for whole, fresh fish remains today, and has even increased in the second half of the 20th Century. The progress carried out in the icing of products and refrigerated transport led to a favorable response to this demand by being able to distribute to any part of the world, in less than five days, a fresh fish of a certified level of hygiene. This tendency is nevertheless affected by the buyers' age and culture, the proximity of the sale and the consumption and the size of the offer. This has benefited over the last 50 years from the development of "ready to eat" food, and from the expansion of large-scale food distribution in the provision circuits. Canned products (tuna, sardines and salted foods), which dominated the market up to the mid-20th Century, are slowly making way for more elaborate products, such as ready meals and frozen foods.

1.4.2. *Diagnosis, fragility and over-exploitation*

Maritime fishing in the Mediterranean is today experiencing a critical situation. It has truly gone beyond the optimal renewal capabilities of most of the exploited stocks. Currently, production from fishing and aquaculture only covers 70% of the food needs of aquatic products of the populations of the countries of the Mediterranean. With a constant increase in the consumption of aquatic products and a stagnation of production, most of these countries find themselves in a greater and greater situation of dependence on imports. This dependence should increase over the next few years, comforted by the necessary reduction of the current levels of exploitation and the development of a highly demanding touristic sector.

Over more than two centuries, the image of an endlessly bountiful sea dominated reflection on the exploitation of its comestible resources. With the exception of words by the likes of "Cassandre" [LEQ 00], the freedom to fish that had followed the freedom of navigation had, since the 17th Century, been met with very few objections. Far greater than the fear of a decrease in resources, it was mainly the fear of overproductions, and therefore a loss of potential profit, that opposed any technological progress.

It was only in the beginning of the 20th Century that the warning bells started to be rung by fishery researchers, resulting in the creation of scientific and diplomatic commissions aiming to limit access to the resource, such as the International Congress for Aquaculture and Fishing at the world exposition of 1900, the International Council for Sea Exploration (ICES) and then in 1914, the *Commission*

Internationale pour l'Exploration Scientifique de la Mer Mediterranée (CIESM).

> Between 1849 and 1850, special Commissions made up of lawyers, sailors and biologists carefully studied the basis of regulation of coastal fishing as, already in this period, the scientific world had recognized the fact that fishing could not be left unchecked on the coasts of France without fearing the rapid destruction of the main species of coastal fish. The scientists of the time had before them the example of the Mediterranean where fishing had been carried out since Antiquity without defined rules, and had caused a general and irremediable decline in the ichthyological fauna. [LED 34]

It is on these terms that the second director of the *Office Scientifique et Technique des Pêches Maritimes* (Scientific and Technical Office of Marine Fisheries), created in 1918, alerts for the scientific communities, professional representatives and politicians to the state of precariousness of the industry and the benefits of providing it with an institutional organization. The scientific desire does, however, go against the socio-economic interests put forward by the states in the name of a necessary growth of the fishing industry to subsidize the growing food and employment needs.

The establishment of true fishing policies by the public powers only took place after World War II, first with the creation in 1945 of the FAO of the United Nations , and then from 1970 the development of the community fishing policies within the EU.

The policies and the strategies will, however, be mainly focused on "development [...] in domains of the reduction of poverty, of food and nutritional security" and more particularly for the EU "the stability of the community market of marine products so as to guarantee quality at reasonable prices and to maintain the income of fishermen at a certain level"[16]. Favoring development and consumption, the states will

16 www.touteleurope.eu/fr/actions/agriculture-peche/politique-commune-de-la-peche/presentation/la-politique-commune-de-la-peche.html.

notably provide financial incentives for the development and modernization of the production tools, while imposing upon the agents of fishing regulatory limitations of use, so as to limit the impact on the resource.

It is only with the observation of the state of over-exploitation of several stocks and new economical and social difficulties of the fishing sector, that the EU adopted in 1992 and 2002 measures aiming to reduce the pressure on the resource, through the reduction of the overcapacity of the European fleet, the limitation of the fishing effort (licenses) and the regulation of the Total Allowable Catches (TACs). Extended to the Mediterranean, the Common Fisheries Policy (CFP) is concretized by the establishment in 2006 of a set of technical conservation methods taking into account the particular context of the Mediterranean and the specificities of the region management systems. So as to minimize the economical and social consequences of these measures, financial aid is guaranteed by the EU (European fisheries fund), the States and the territorial collectivities. Not theoretically aimed at increasing the fishing capacities, this aid has favored the modernizing of equipment, and notably communication and detection. They have contributed to the improvement of security, hygiene and working conditions. They have, therefore, actually contributed to the improvement of the effectiveness of fishing vessels. The situation is even more critical for countries of the south coast of the Mediterranean, which, undergoing high levels of demographic growth and with a commercial deficit, have based part of their hope of development on the modernization of their fishing fleet and the connected industries. The fishing policies are, therefore, more than even faced with the difficulty of combining economical and social development with the conservation of resources.

1.4.3. *State of the resource and the environment*

The overcapacity of fleets, an insufficient mastery of the management of the stocks exploited and erroneous fishing strategies have caused takings exceeding the reproductive capabilities of the species, which has in turn resulted in the decrease in the abundance, size and shapes[17] and a reduction of the specific components of the catches.

Without omitting the effects of urbanization, touristic development and water pollution, fishing is also held responsible for the modification of the environments exploited and the survival of vulnerable species.

The causes of the decline of these species are likely due to the combination of direct human impacts that have increased in intensity during the 20th Century and the intrinsic characteristics of this fauna, such as their slow growth, their longevity, their relatively low fecundity and their position as trophy catches. The impact of anthropic factors, such as global warming or the modification of waterways, the acidification of the oceans and increases in salinity play an aggravating role that is, however, difficult to evaluate [GRO 14, MAY 11].

It is only with the first international congress for the protections of nature (Paris, 1923), and the convention on the conservation of natural state fauna and flora (London 1933) that for the first time the notions of "species threatened with extinction", "integral natural reserves" and "national parks" were covered [FRO 98]. However, it was only the creation of the International Union for the Conservation of Nature (IUCN), in Fontainebleau, in 1948, which marked the start of modern engagement of the international community in favor of the protection of nature. The awareness of the tight links between the exploited resources and the environment became especially important at the beginning of the 1990s with the creation of work groups in the international commission for the exploration of the sea on the effects of fishing on ecosystems [HAL 99].

1.4.4. *Ecosystem approach of fisheries and governance*

The Ecosystem Approach of Fisheries (EAFs), appearing with the Code of Conduct for Responsible Fisheries (FAO, 1995), arose from

17 Fishing techniques, such as trawling, gillnetting and trap fishing that use meshes, favor the escape of fusiform and oblong species at the expense of more angular-shaped species or those with extensive skin expansions.

the observation of the dependence of the exploited species on each other and their environment, and the realization that the food provided by the fisheries industry is only one of the services provided by the environment, such as functions of the regulation of the climate, energy production, the provision of cultural or touristic activity [GAS 09].

The sustainable use of marine species can, therefore, only be ensured by measures combining ecological and socio-economic requirements and established through management plans defined specifically for the ecosystems concerned. By considering that man and his activities are an integral part of the ecosystem, this mode of governance imposes that benefiting parties are all associated with all of the stages of the process of implementing and applying these plans.

The application of this particular mode of governance aims to reduce the impact of human activity on all of the behaviors of the ecosystem. However, it does not have to involve the detailed knowledge of all of the interactions before taking decisions aiming to prevent deterioration of the environment.

On this basis, the General Fisheries Commission for the Mediterranean (GFCM) following its 36th session (Marrakesh, 14–19 May 2012) adopted the principle of the application of multiannual management plans chosen case-by-case, according to the scientific recommendations of its Scientific Advisory Committee (SAC).

The multiannual management plans must be designed to counter and prevent overfishing, while providing high, sustainable yields, and maintaining, as far as possible, the size of the stocks of the species exploited at levels that provide the highest sustainable yield possible, and with the smallest risk of a reduction of the stock beyond safe biological limits.

With this recommendation, the GFCM is reaffirming its desire for a spatial management of the resources, through large maritime areas (*Geographic Spatial Area*) and its desire to take into account the diversity of the socio-economic context of each country.

Figure 1.14. *Sea bass and gilt-head sea bream farms in Greece (Argolida Gulf, 2013) (source: Daniel Faget)*

COMMENTS ON FIGURE 1.14.– Presented as an alternative to fishing by capture, aquaculture farms are in full development in the Mediterranean. Greece appears as one of the leading countries in this domain, since its production of marine species (111,000 tons) in 2008 equaled 30% of the total Mediterranean production (376,000 tons).

1.4.5. *Institutions*

In virtue of the United Nations Convention on the Law of the Sea (UNCLOS), the states bordering the Mediterranean must cooperate with each other, either directly or through an appropriate regional organization so as to, notably, coordinate the management, conservation, exploration and exploitation of the biological resources of the sea. To this day, two regional organizations dealing with the problems of sea fishing have been established: the ICCAT and the GFCM.

ICCAT, established in 1969, has the objective of guaranteeing the sustainable exploitation of tuna in the Atlantic Ocean and adjacent waters and, as a result, in the Mediterranean. The commission can, on the basis of the scientific data available, make recommendations aiming to maintain tuna populations at levels that allow them to be exploited.

Created in 1952 under the FAO, the GFCM plays a major role in the management of fisheries in the Mediterranean and the Black Sea. Its 23 member states have had the power, since 2004, to adopt the restrictive recommendations submitted by a scientific committee, for the conservation of resources and the management of fisheries in the area covered by the Convention. A Committee of Compliance is charged in particular with examining the application of the measures adopted and eventually making recommendations for their application.

In collaboration with these two main organizations, various other international institutions contribute to the promotion in the Mediterranean of the modes of development integrating conservation and sustainable development, such as the Mediterranean Action Plan (MAP), the IUCN or the Agreement on the Conservation of Cetacean of the Black Sea, Mediterranean Sea and Contiguous Atlantic Area (ACCOBAMS).

1.4.6. *Seeking to promote the value of seas*

Faced with the precariousness of the resources, the hopes are very soon led to possibilities of increasing the possibility of the environment, by adapting spaces for farming (valliculture and aquaculture) or the protection of juveniles and reproducing individuals (artificial reefs, *cantonments*, reserves, etc.).

1.4.6.1. *Aquaculture*

Like in the 19th Century, aquaculture appears as the only possible solution for the dilemma caused by the need to respond to the increasing demand for aquatic products, and the obligation to reduce the amount of fish caught. Its main development came from the principle of the fattening of captured fish (*capture-based aquaculture*). The best-known examples in the Mediterranean are the eel, mullet and, more recently, bluefin tuna [LOV 08]. It is only around the 1980s that aquaculture really started to take off spectacularly with the mastering of the reproduction of species of high-commercial value such as bass and sea bream. The farming of red tuna is thus presented as an ideal candidate in aquaculture of the future, notably due to the rapid growth of young individuals [LAB 10]. The development of aquaculture does, however, remain faced with the need to decrease its dependence on the resource, notably in the feeding of farms and its impact on surrounding areas.

Like industrial fishing, aquaculture has imposed itself on the professionals of neighboring artisanal fisheries, taking away part of their traditional activity and access to the market.

Its interactions with other human activities, such as the environment, are a result of the major questions. The Mediterranean countries decided to take them seriously into consideration, due to their socio-economic importance, and due to the growing importance of aquaculture for the food provision of main neighboring countries (Resolution CGPM/36/2012).

1.4.6.2. *Protected marine areas*

The interest in developing new methods of management based on protected marine areas (PMAs) has been revived in the last decade, supported by the feeling that it is possible to simultaneously protect threatened species and ensure the practice of certain activities of commercial fishing. For this, the PMAs must adapt their observation system to new objectives and integrate into their management the managing of all of human activity; this necessarily implies the participation of all parties – professional, scientific and administrative.

Several scientists and managers consider that the existing PMAs are often too small, and the status of the species and the habitats that they are meant to protect depend not only on the pressure of external fishing but also on the migration of the species. The creation of larger protection zones, as authorized by the UN convention on the rights of the sea (Montego Bay, 1994), would allow for a more effective application of the protective measures by better taking into account the scope of activity of the different maritime activities.

Another major difficulty is the management of PMA created beyond territorial waters, in "open" sea. In this space, which represents 60% of the surface of the Mediterranean waters, the Law of Seas covers surface navigation and the exploitation of mineral resources, while fishing is covered by the GFCM and threatened species by the Barcelona convention. The application of the measures necessary for scientific observation and control would involve the creation of a single international jurisdiction covering all of human activity [SAC 08].

1.4.7. *Education, raising awareness and labeling: the fishermen, agents of a sustainable exploitation of the environment*

One of the key factors for a sustainable exploitation of the environment is the integration of the fishing industry into the process of the management of the ecosystems exploited and in the choice of measures of application. A non-negligible part of the policy and regulation of fisheries is currently defined by the professionals themselves, as part of their respective instances. Among these, the Regional Advisory Council of the Mediterranean (RAC MED), created in 2008 as part of the common fisheries policy, is composed of representatives of fishing and other relevant groups such as organization for the protection of the environment, the consumers, the representatives of recreational or sport fishing. Its role is to allow the European commission to benefit from the knowledge and experience of the sector in the formulation and application of measure of fisheries management.

The awareness of the limited and precariousness of the resource has led to a questioning of the principle of a development of fishing essentially established on a greater and greater productivity of the fishing system. This economic system, born in the industrial revolution, became prominent in an era when it appeared to be the most efficient solution to ensure adequate provision of food to the population, and the social well-being of fishermen. The fishing industry must replace this conception of progress with a new approach, first of all founded on the preservation of the exploited ecosystems. The adhesion of those involved in this new form of management implies a clear demonstration of the potential benefit that companies will be able to experience. Any development project or conservation project must necessarily plan an evaluation of the biological and socio-economic consequences of its impact. Raising the awareness of the agents and consumers to this new concept could also be stimulated by campaigns of eco-labeling (or eco-tagging) in support of practices of sustainable fishing. In this spirit, the fisheries and transformation companies must be able to prove the sustainability of their activity to obtain an eco-label. This approach does, however, require the establishment of reliable references that are not based on market considerations. In this aim, the FAO, during the 26th session of its Committee on Fisheries (FAO-COFI, Rome, 2005), exposed the general principles that must define eco-tagging, notably the "necessity for independent and reliable auditing systems, of transparency in the establishment of norms and of the sharing of responsibility, as well as the rules with solid scientific bases". In this spirit, the EU and France in particular have since started initiatives to determine the regulatory conditions of the eco-labeling of marine products.

1.4.8. *The necessary consideration of the patrimonial dimension of artisanal fishing*

Relatively neglected in fishing policy established around the Mediterranean in the second half of the 20th Century, coastal fishing fleets are today the objects of reflection linked to the definition of sustainable fishing. A living natural environment does require living fishing communities, which guarantee, beyond their traditional role of supplying the markets, the maintenance of local cultures. This patrimonial dimension appears today as constitutive of the rich identity of the Mediterranean coasts. It is today highly threatened. Its preservation is vital for, in the next decade, the reorientation of touristic policy in the Mediterranean, more open to the promotion of "blue tourism", on the model of "green tourism", very popular today in the countryside of occidental Europe. Guardians of landscape and knowledge, the agents of small-scale fishing remain today the first guarantors of the quality of their areas of activity. Upholding daily relations with the marine environment, they are able to detect the slightest of change. They must, therefore, today become more than ever essential partners in the application of the management and preservation policies for fisheries. The preservation of these fisheries is, therefore, not reduced to a simple choice between destruction of the resource and making a sanctuary of this environment. The experiments carried out over 20 years around the national park of Port-Cros show that by favoring artisanal fishing over recreational fishing, and by implementing strict regulation over the practice, the density and the biomass of the fish remain high. The abandoning of certain techniques, such as the banning of trawling to the benefit of controlled trammel fishing, finally leads to catches by units of effort equal to those of the non-protected surrounding zones [BOU 04, PRO 13].

1.5. Bibliography

[ALE 03] ALEGRET I., TEJERO J.L., MARTI I., *et al.*, *La pesca a Catalunya*, Angle Editorial, Barcelona, 2003.

[AMO 60] AMOUROUX A., "Valras, étude d'une station touristique du Languedoc Méditerranéen", *Méditerranée*, no. 1, pp. 67–91, 1960.

[APP 13] APPERT F.N., *Le livre de tous les ménages, ou l'Art de conserver pendant plusieurs années, toutes les substances animales et végétales*, 3rd ed., Barrois l'aîné, Paris, 1813.

[BAR 61] BARBAZA Y., "La pêche méditerranéenne: mise au point", *Annales de Géographie*, vol. 70, no. 377, pp. 60–70, 1961.

[BAS 55] BAS C., MORALES E., RUBIO M., La pesca en España. I Cataluña, Consejo Superior de Investigaciones Científicas, Fisheries Research Institute, Barcelona, 1955.

[BAS 05] BAS PEIRED C., The Mediterranean sea: living resources and exploitation, CIHEAM – IAMZ/FAO COPEMED Project, 2005.

[BER 87] BERGER A., FERAL F., "L'interventionnisme de la puissance publique dans le domaine des ressources marines vivantes", *Revue d'Economie Maritime*, vol. 35, no. 139, pp. 3–31, 1987.

[BER 98] BERTRAND R., *Le Vieux-Port de Marseille*, Jeanne Laffitte, Marseille, 1998.

[BOU 87] BOUDE J.P., "Les pêches méditerranéennes", *Revue études internationales*, vol. 18, no. 1, pp. 83–105, 1987.

[BOU 04] BOUDOURESQUE C.F., CADIOU G., GUERIN B., *et al.*, Is there a negative interaction between biodiversity conservation and artisanal fishing in a marine protected area, the Port-Cros National Park (France, Mediterranean Sea), Scientific Reports of Port-Cros National Park, France, no. 20, pp. 147–160, 2004.

[BUT 98] BUTI G., "Madragues et pêcheurs provençaux dans les mailles des pouvoirs (XVIIe-XIXe siècles)", in LE BOUEDEC G., CHAPPÉ F. (eds.), *Pouvoirs et littoraux du XVe au XXe siècle, Actes du colloque international de Lorient,* Rennes University Press, Rennes, pp. 57–73, 24–26 September 1998.

[BUT 00] BUTI G., "Résonances urbaines des conflits de pêche en Provence (XVIIe-XIXe siècles)", *Provence historique*, vol. L, no. 202, pp. 438–457, October–December 2000.

[BUT 04] BUTI G., "Gens de mer et du terroir: capitaines – vignerons et marins forestiers de la France méditerranéenne au XVIIIe siècle", in LE BOUEDEC G., PLOUX F., CÉRINO C., *et al.* (eds.), *Entre terre et mer. Sociétés littorales et pluriactivités (XVe-XXe siècles)*, Rennes University Press, Rennes, pp. 147–185, 2004.

[CAB 95] CABANTOUS A., *Les citoyens du large. Les identités maritimes en France (XVIIe-XIXe siècles)*, Aubier, Paris, 1995.

[CEN 01] CENTI C., TEMPIER E., Eléments de prospective sur la pêche et les cultures marines. Les leçons du cas Méditerranéen, Centre for the Study of Social Relations, University of the Mediterranean, 2001.

[CIC 06] CICTA, Rapport de la 15e réunion extraordinaire de la Commission (Dubrovnik, Croatia, 17–26 November 2006), Rapport de la période biennale, 1re partie (2006), vol. 1, Commission Internationale pour la Conservation des Thonidés Atlantiques, Madrid, 2006.

[DES 09] DESCLERS S., "European fishing fleet capacity management", *Seas at Risk. Background Paper, Abstracts and Conference Proceedings*, *Seas at Risk Annual Conference*, Brussels, Belgium, pp. 21–22, 21 October 2009.

[DIE 53] DIEUZEIDE R., NOVELLA M., *Le Matériel de pêche maritime utilisé en Algérie. Fibres et fils, cordages et câbles, filets et engins*, Imprimerie Imbert, Alger, 1953.

[DUQ 00] DUQUESNE C., “De la convergence européenne à la mise en œuvre différenciée: la politique commune de la pêche”, *Colloque CEVIPOF – CERI L'intégration européenne entre émergence institutionnelle et recomposition de l'Etat*, Sciences PO, Paris, pp. 2–35, 26–27 May 2000.

[ETI 10] ETIEVANT P., BELLISLE F., DALLONGEVILLE J., *et al.* (eds.), Les comportements alimentaires. Quels en sont les déterminants? Quelles actions, pour quels effets?, Expertise scientifique collective, report, INRA, France, 2010.

[EUR 02] EUROPEAN FOUNDATION FOR THE IMPROVEMENT OF LIVING AND WORKING CONDITIONS, ICT driving change in the fishing industry: a review of the years 1990-2002, 2002.

[EVO 11] EVOMED, The 20th Century evolution of Mediterranean exploited demersal resources under increasing fishing disturbance and environmental change, EVOMED, final report January 2011, Contract EU DGMARE SI2 539097, 2011.

[FAG 05] FAGET D., “La découverte d'un risque sanitaire lié à l'alimentation: fièvre typhoïde et consommation de coquillages en Méditerranée du milieu du XIXe siècle à la Seconde Guerre Mondiale”, *Provence historique*, vol. LV, no. 221, pp. 345–365, July–September 2005.

[FAG 07] FAGET D., “Cultiver la mer: biodiversité marine et développement de l'ostréiculture dans le Midi méditerranéen français au XIXe siècle”, *Annales du Midi*, vol. 119, no. 258, pp. 207–226, April–June 2007.

[FAG 11] FAGET D., *Marseille et la mer. Hommes et environnement marin (XVIIIe-XXe siècles)*, Rennes University Press/Publications of the University of Provence, Rennes/Aix-en-Provence, 2011.

[FAO 03] FAO, DÉPARTEMENT DES PÊCHES, “Aménagement des pêches. 2. L'approche écosystémique des pêches”, *FAO Directives techniques pour une pêche responsable*, no. 4, Suppl. 2, 2003.

[FER 81] FERRETTI M., Evolution de la technologie dans la pêche au chalut pélagique en Adriatique, FAO Fisheries Report no. 253, pp. 159–161, 1981.

[FER 01] FERAL F., Sociétés maritimes, droits et institutions des pêches en Méditerranée occidentale. Revue synthétique des droits collectifs et des systèmes décentralisés de discipline professionnelle, FAO Fisheries Technical Paper No. 420, 2001.

[FOL 93] FOLSOM W.B., ROVINSKY D.J., WEIDNER D.M., "Western Europe and Canada [fishing fleets]", *World Fishing Fleets: An Analysis of Distant-water Fleet Operations. Past-Present-Future*, Prepared by the Office of International Affairs, National Marine Fisheries Service, NOAA, U.S. Department of Commerce, Silver Spring, MD, vol. 6, pp. 1–362, November 1993.

[FRO 98] FROMAGEAU J., *Du droit transfrontalier à la protection de la biosphère*, KISS A., BEURIER J.P. (eds.), Frison-Roche, Paris, 2000.

[GAR 11] GARCIA S.M., *Long-Term Trends in Small Pelagic and Bottom Fisheries in the Mediterranean: 1950-2008*, Plan Bleu, Valbonne, 2011.

[GAS 09] GASCUEL D., "L'approche écosystémique des pêches, une condition pour l'exploitation durable des océans", *POUR*, Edition du GREP, Paris, no. 202–203, pp. 199–206, 2009.

[GIO 95] GIOVANNONI V., *Les pêcheurs de l'étang de Thau, écologie humaine et ethnologie des techniques*, L'Harmattan, Paris, 1995.

[GOU 91] GOURRET P., "La Consommation et le Commerce du Poisson, des Coquillages, etc., à Marseille", *Annales du Musée de Marseille zoologie*, vol. 4, no. 1, p. 20, 1891.

[GRO 14] GROS P., PROUZET P., "Impact of global change on the dynamics of marine living resources", in MONACO A., PROUZET P. (eds.), *Ecosystem Sustainability and Global Change*, Chapter 4, ISTE, London, John Wiley & Sons, New York, pp. 49–130, 2014.

[HAL 99] HALL S.J., *The Effects of Fishing on Marine Ecosystems and Communities*, Blackwell Science, Oxford, 1999.

[HUR 02] HURTADO V., MESTRE J., MISERACHS T., *Atles d'història de Catalunyia*, Edicions 62, Barcelona, 2002.

[LAB 10] LABAT J., COUTARD S., PLANTAMP C., *et al.*, "Etat des lieux et perspectives de la filière thon rouge en Méditerranée", *Document Etudes Agro-Campus*, Rennes, 2010.

[LAR 97] LARGUIER G., "Pêche et société en Languedoc au XVIIIe siècle", *L'histoire grande ouverte. Hommages à Emmanuel Le Roy Ladurie*, Fayard, Paris, 1997.

[LED 34] LE DANOIS E., "Rapport sur les mesures règlementaires à prendre d'urgence pour remédier à la situation critique de l'industrie de la pêche due à l'appauvrissement des fonds", *Revue de l'Office des Pêches Maritimes*, vol. 8, no. 11, p. 24, 1934.

[LED 35] LE DANOIS E., (ed.), "Manuel des Pêches maritimes françaises", *Mémoires de l'Office des Pêches Maritimes* (special series), vol. 10, no. 2, p. 199, 1935.

[LEQ 00] LEQUESNE C., "De la convergence européenne à la mise en œuvre différenciée: la politique commune de la pêche", *Colloque CEVIPOF – CERI L'intégration européenne entre émergence institutionnelle et recomposition de l'Etat*, Sciences PO, Paris, pp. 1–35, 126–27 May 2000.

[LLE 98] LLEONART J., LLORET J., TOUZEAU S., *et al.*, "Mediterranean fisheries, an overview" (reviewed by FROMENTIN J.M., LEVI D., STERGIOU K.I., *et al.*), *II SAP Meeting*, Barcelona, Spain, 13–17 October 1998.

[MAR 91] MARION F., "Note sur Exploitation méthodique des fonds maritimes littoraux", *Annales du Musée d'Histoire naturelle de Marseille. Travaux de Zoologie appliquée*, Marseille, série I, vol. 4, pp. 2–5, 1891.

[MAR 05] MARTY F., *Entre mailles et filets. Savoir-faire des pêcheurs du Var*, Jeanne Laffitte, Marseille, 2005.

[MAU 61] MAURIN C., DI MEGLIO S., "Evolution de la pêche à la Sardine sur les côtes françaises", *Science & Pêche*, vol. 98, pp. 1–8, 1961.

[MAY 11] MAYNOU F., SBRANA M., SARTOR P., *et al.*, "Estimating trends of population decline in long-lived marine species in the Mediterranean sea based on fishers", *Perceptions*, *PLoS ONE*, vol. 6, no. 7, p. 10, 2011.

[MEN 69] MENC A., "Quelles sont les causes de la diminution de la pêche sur les côtes de la Provence? Et les moyens de la rendre plus abondante?", *Recueils des pièces de poésie et d'éloquence présentées à l'Académie des Belles-lettres*, Sciences et Arts de Marseille pour les prix des années 1768 et 1769, Sibié, Marseille, Imprimeur du Roi, 1769.

[MER 63] MERLIN P., "Les pêcheurs de Mers-el-Kébir", *Méditerranée*, no. 4, pp. 27–54, 1963.

[MEU 87] MEURIOT E., DREMIERE P.Y., CAPELLE J., "Le chalutage en Méditerranée: Le port de Sète", *Rapports économiques et juridiques de l'IFREMER*, no. 3, p. 147, 1987.

[MON 14a] MONACO A., PROUZET P. (eds.), *The Land-Sea Interactions*, ISTE, London, and John Wiley & Sons, New York, 2014.

[MON 14b] MONACO A., PROUZET P. (eds.), *Ecosystem Sustainability and Global Change*, ISTE, London, and John Wiley & Sons, New York, 2014.

[MUS 21] MUSSET R., “Le port de Lorient”, *Annales de Géographie*, vol. 30, no. 166, pp. 310–312, 1921.

[PAY 07] PAYN-ECHALIER P., *Les marins d’Arles à l’époque moderne*, Publications of the University of Provence, Aix-en-Provence, 2007.

[PER 01] PERARD J., MAIRE M., Congrès international d’aquiculture et de pêche: mémoires et comptes rendus des séances par M.J. Pérard et M. Maire, Ministry of Commerce, Industry, Posts and Telegraphs, World Expo 1900, Paris, A. Challamel, 1901.

[PIO 08] PIOCH S., Les habitats artificiels: éléments de stratégie pour une gestion intégrée des zones côtières? Essai méthodologique d’aménagement en récifs artificiels adaptés à la pêche artisanale côtière, Thesis, EGIS EAU, 2008.

[PRO 13] PROUZET P. (ed.), Prise en compte de la dimension patrimoniale dans la définition de la durabilité des modes d’exploitation des ressources halieutiques, Programme REPERE, MEDD, 2013. Available at www.programme-repere.fr/.

[RAV 03] RAVIER-MAILLY C., Fluctuations à long terme du thon rouge: validité, origines et conséquences, Thesis, National School of Agronomy of Rennes, 2003.

[RIO 02] RIOS JIMENEZ S., “La industrialización de la pesca en la provincia de Huelva [1800-1930]”, *Historia Agraria*, pp. 45–67, no. 28, December 2002.

[SAC 08] SACCHI J., “Impact des techniques de pêche sur l’environnement en Méditerranée”, *Etudes et revues. Commission générale des pêches pour la Méditerranée*, FAO, Rome, no. 84, p. 62, 2008.

[SAC 11] SACCHI J., *Analyse des activités économiques en Méditerranée: secteurs pêche – aquaculture*, Plan Bleu, Valbonne, p. 87, 2011.

[SIM 61] SIMONNET R., “Essai sur l’économie des pêches maritimes en Algérie”, *Revue des Travaux de l’Institut des Pêches Maritimes*, vol. 25, no. 1, pp. 33–124, 1961.

[SUE 45] SUE E., *Le commandeur de Malte*, Gosselin, Paris, 1845.

2

Microalgae and Biotechnology

2.1. Microalgae

Microalgae are a group of heterogeneous, unicellular, photosynthetic, eukaryotic organisms[1]. The first event of endosymbiosis, at the origin of all photosynthetic eukaryotes, goes back to 1.8 billion years ago [FIN 10], while terrestrial upright plants only diverged from green algae 500 million years ago [KEE 99]. The photosynthetic yield of microalgae is slightly greater than that of terrestrial plants [WIJ 10] and the fact that they evolve in an aqueous medium, which gives them direct access to their nutritional elements, explains in part a better productivity and a higher growth.

Depending on the ambient conditions, such as salinity, light, concentrations of nutrients, etc., the size and appearance of microalgae can change greatly, making their identification difficult. A recent article [GUI 12] suggests a reasonable number of 72,000 estimated species. We can also find estimations going up to 40,000 and 60,000 known species, and assumptions on the number of non-described

---

Chapter written by Jean-Paul CADORET, Gaël BOUGARAN, Jean-Baptiste BÉRARD, Grégory CARRIER, Aurélie CHARRIER, Noémie COULOMBIER, Matthieu GARNIER, Raymond KAAS, Loïc LE DÉAN, Ewa LUKOMSKA, Elodie NICOLAU, Catherine ROUXEL, Bruno SAINT-JEAN and Nathalie SCHREIBER.

1 Photosynthetic prokaryotes such as *Prochlorococcus* and *Synechococcus* are of major importance at a global level, and are considered to be key actors. However, it would be ambitious to attempt to cover the entire photosynthetic aquatic world. This chapter is essentially oriented toward microalgae and therefore eukaryotes.

species is in the order of hundreds of thousands of species [ROS 10]. Microalgae are distributed over the surface of the globe, whether in marine, freshwater or brackish water environments. They have colonized all mediums, from polar ice to deserts or hot water sources. They have adapted themselves to extreme environments, living in salt marshes, acidic media, or even in very low-light conditions. They contribute toward 90% of primary aquatic production and 50% of global primary production (Figure 2.1). They are at the bottom of the aquatic food chain. The annual ocean production is estimated to be at 100×10^9 tons of dry matter [PAU 95]. Through their presence on the surface of the oceans (covering 70% of the planet), they play a major role in the global climate by turning CO_2 into organic matter [RAV 99].

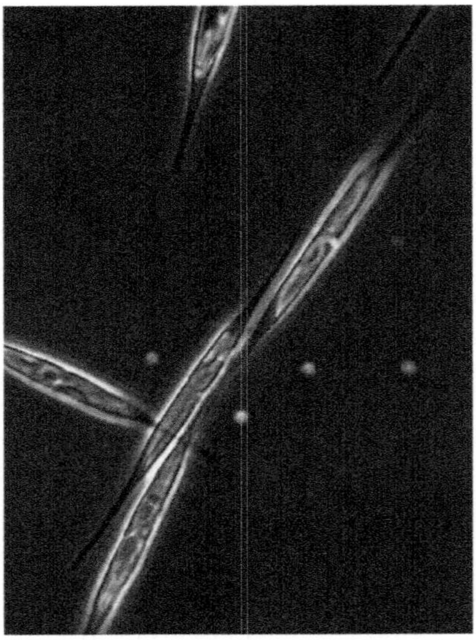

Figure 2.1. *Cylindrotheca closterium, a benthic pennate diatom over around 100 µm in length (photo Kaas/Ifremer)*

This biological diversity, a result of excellent adaptability, suggests a proportional richness in unique molecules and therefore of applications in the field of biotechnologies. Twenty species of microalgae are the subject of valorization for providing carotenoids, fatty acids and other polysaccharides. The systematic screening of known species and, in the future, of species to be discovered, will benefit from technological advances, allowing the low-cost sequencing of genomes and screening automation that Genopole or other private consortiums are becoming equipped with. A complementary path in this research for original molecules consists of transferring the genes of chosen molecules by genetic engineering and profiting from all the advantages offered by unicellular microalgae.

Rising petrol prices have propelled microalgae forward as a source of vegetal biomass in the same league as terrestrial plants. Indeed, phytoplankton will play two major roles in the near future, not only as an energy source but also as a source of precursor primary matter for polymers in the domain known as "green chemistry". Microalgae present singular characteristics that provide advantages over terrestrial plants. The following section describes these differences which are, for example, better growth, better yields per acre due to superior photosynthetic activity. In terms of energy, microalgae thus accede to the rank of third generation.

2.2. The potential value of microalgae

2.2.1. *Human nutrition*

Certain Amerindian and African populations have long included spirulina in their diets, and insist not only on the nutritional values of these cyanobacteria, but also on their preventative capabilities against certain pathologies [JEN 01]. Without counting macroalgae (for which 19 species can be consumed as vegetables or condiments), the current market is dominated by spirulina and chlorella, and the only diatom to have received this approval in France is *Odontella aurita* (Figure 2.2). At a global level, we can add the green microalgae *Dunalliela salina* for its ß-carotene. Biscuits made from the microalgae *Tisochrysis lutea*, rich in omega-3, have also been made [GOU 08]. A large amount of investigation remains to be done for teams with enough energy to homologate other microalgae as foods [JEN 01]. However, the production costs do greatly hinder the applications in this domain [BEC 07].

2.2.2. *Animal nutrition*

Microalgae represent an important source of food in animal production, outside of aquaculture. An estimated 30% of the global algae production is for the feeding of terrestrial animals, representing a huge market. The most common species in this market are spirula, chlorella and *Scenedesmus* sp.

However, one of the first fields of application where algae found a place was in aquaculture. The production of phytoplankton remains an obligatory step in the farming of mollusks, whichever their stage of development, particularly in the larvae stages, for which the food source must be living. While the plethora of microalgae concerned represents some tens of species, only a dozen are commonly cultivated. Scientific studies showing the links between the choice of the cocktail of species making up the algal food and the quality of the farms still have not been carried out [RIC 06].

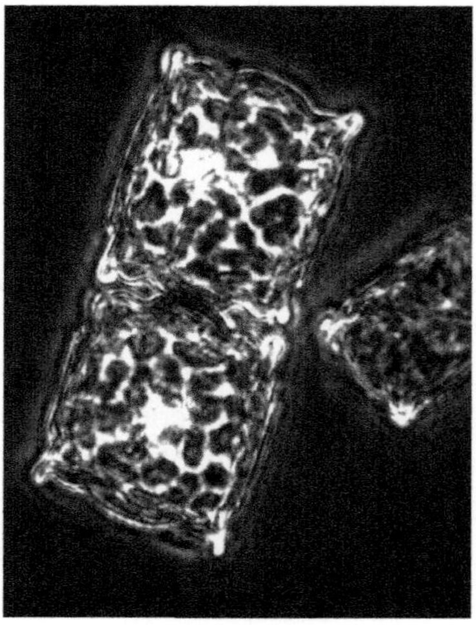

Figure 2.2. *Odontella aurita: length of around 50 μm; this large diatom is regularly cultivated in Vendée (photo Kaas/Ifremer)*

Microalgae are also used to raise the first larval stages of some fish, but they could also represent an alternative to using fishmeal as feed for fish farming. In 2006, the aquaculture sector consumed a total of 62% of the global production of fishmeal (or more than 32 million tons) and 88.5% of the global fish oil production, or 835,000 tons [TAC 08]. In the context of the preservation of biodiversity and the problem of managing stocks of marine species, the total or partial replacement of fishmeal and fish oils by microalgae is of great economic and ecological importance. Thus, generally, the replacement of fish oils by algae is perfectly foreseeable [ABU 09], even if the economic dimension should not be underestimated. And, if it is possible to feed bass with dried microalgae [TUL 12], we can also incorporate 5–10% of microalgae in chicken feed, with some effect on the coloration of the meat and of the yolk [BEC 07].

2.2.3. *Health*

2.2.3.1. *Fatty acids*

Polyunsaturated fatty acids with a gamma-linolenic acid (GLA), arachidonic acid (AA), eicosapentaenoic acid (EPA) or long chain docosahexaenoic acid (DHA), produced by microalgae, accumulate in most marine animals. Consumption of sufficient quantities of this type of fatty acid would have a positive impact on health: protective properties against cardiovascular diseases and cancers. The oil extracted for example from the microalgae *Schizochystrium* sp. (authorized as a food product) contains 35–45% of DHA. In comparison, the traditional oils richest in omega-3 (nut oil, colza oil, etc.) contain around 10% of alpha-linolenic acid, a precursor of the omega-3 family. The main known sources of omega-3 in algae are *Crypthecodinium cohnii* and *Schizochytrium sp.*

2.2.3.2. *Pigments*

Since the establishment of Recording, Evaluation and Authorization of Chemicals (REACH), entrepreneurs have been led to take an interest in the replacement of synthetic molecules by molecules coming from resources said to be sustainable, such as pigments, which can intervene in the domain of human feeding as natural colorings.

Microalgae contain three big types of photosynthetic pigments: the phycobiliproteins, chlorophylls and carotenoids.

Phycobiliproteins are pigments mainly present in cyanobacteria, red algae and glaucophytes. They are water-soluble macromolecules, whose colors are different: red for phycoerythrin and blue for phycocyanine and allophycocyanine. These colorings can be used in the agro-alimentary industry to color food and drink products. Phycocyanine is the only natural blue coloring authorized in Europe. The natural products present the advantage of decreasing risks of allergens [MOR 12].

Chlorophyll is present in all photosynthetic organisms. It is mainly used as green food coloring (E 140). As a photosensitive, or "photosensitizing", pigment, chlorophyll can transfer its excitation energy toward its molecular environment to create reactive oxygen species (ROS). These molecules, which are highly reactive, initiate powerful oxidation mechanisms, whose effects at high doses can be noxious for a cell [ROB 09]. As a promoter of reactive oxidation chains, which can be controlled, chlorophyll and its derivatives allow for varied application: anti-bacterial action, scarring of tissue, studies of the oxidation of materials, oxidative cellular stress, etc. It was shown early on that it accelerates scarring of the skin [HOR 51], stimulates the growth of tissues and is used in the treatment of ulcers by reducing pain and improving the aspect of damaged tissue [CAD 48].

Carotenoids are the pigments with the greatest diversity among natural products. In 2004, more than 700 had been described [BRI 04]. The carotenoids are made of an unsaturated carbon chain (like α- and ß-carotene), while the xanthophylls (such as astaxanthine, luteine and zeaxanthine) are oxygenated derivatives. Most of the carotenoids are powerful antioxidants [CHR 13]. We will mention ß-carotene, referenced as E160a, whose rate in spirula is 10 times greater than in the carrot, astaxanthine (E161j) used in the farming and coloring of salmon which is the only pigment fixed by the muscles of farm salmon, lutein (E161b) and zeaxanthin (E161h) added to foods to color the skin of chickens and increase the yellow color of yolks. Outside of their use in food, carotenoids also have potential applications in human health through their ability to isolate reactive oxygen species. Beta-carotene, or pro-vitamin A is a vital co-factor in preventative child health. Epidemiological studies have shown that intake of carotenoids decreases the prevalence of cancers and inflammatory diseases [GAG 12], particularly lycopene. Beta-carotene, astaxanthin, cantaxanthin and zeaxanthin all present anti-tumor activity in the liver. Lutein and zeaxanthin are the pigments responsible for the maintenance of normal vision in humans. These pigments are added as supplements in the diet of persons suffering from diseases of the retina, such as age-related macular degeneration (ARMD). Preventatively, carotenoids protect the skin from UV rays and are used with other anti-oxidants in sun protection products (see *infra*) [VIL 11], suggesting a wide field of application for microalgae.

2.2.3.3. *Polysaccharides*

These polymers, or rather, their sulfated forms, have antiviral and anti-proliferative capabilities, revealed in lines of cancer cells, but also in rabbits during *in vivo* experiments. They are already extracted from chlorella and from the red microalgae *Porphyridium purpureum*. These compounds could appear in the formulation of hydrogels for bone reconstruction and in the vectorization of osteo-inducing vectors. Exopolysaccharides (EPS) taken from microalgae and their applications are reviewed by [RAP 13].

2.2.3.4. *Antioxidants*

The assimilation of oxygen by an organism can lead to the creation of dangerous derivatives, including singlet oxygen and free radicals. These highly reactive forms of oxygen play an important role in various chronic (cancer atherosclerosis, arthritis, Parkinson's, etc.) or acute (inflammation, etc.) pathologies. Like antioxidants, the superoxide dismutases (SODs) are of interest in protecting tissues. This SOD is one of three catalysts involved in the conversion of the superoxide anion O_2 into H_2O_2. The sequence of this metalloprotein (there are three classes: Cu/Zn, Mn or Fe) is known. SODs are not, however, the only antioxidants used. In plants we can cite ascorbate peroxidase, monodehydroascorbate reductase, glutathione reductase, ascorbic acid, glutathione and the phenylpropanoid pathway. It is in chlorella that [KO 12] describe the anti-oxidant properties of a peptide.

2.2.3.5. *Cell factory*

The controlled production of simple and complex molecules for industrial, nutraceutic and therapeutic use represents an extraordinary opportunity for the development of a market quickly moving forward (market estimated at several tens of billions of dollars) [SCH 04]. There are currently many systems for the production of recombinant proteins (bacteria, yeast, mammalian cells and plants) which allow the production of more or less complex molecules such as insulin, growth hormones, monoclonal antibodies, as well as other therapeutic proteins. However, all of these systems present advantages as well as several disadvantages, such as production costs, regulation in terms of sanitary security (dissemination of genetically modified organisms (GMOs), absence of pathogens such as viruses, prions), ease of extraction and purification of the proteins chosen. Apart from the growing interest in microalgae as natural sources of high value compounds (pigments, polyunsaturated fatty acids, polysaccharides, etc.), they also present many advantages as factories for the production of recombinant proteins used therapeutically [CAD 08]. This promising technology has an objective for the simplification of production processes and the reduction of production costs. Compared with the current systems for the production of recombinant proteins, microalgae present many competitive advantages. Indeed, most microalgae are photoautotrophic organisms, depending only on water, light and base nutrients such as nitrogen and phosphorus and carbon present in the form of carbon dioxide. These culture conditions are therefore much cheaper than the media used notably for mammalian cells. Finally, microalgae present high productivity levels (close to those of bacteria). The possibility of culturing them in confined and controlled media gets rid of the problems linked with the dissemination of GMOs into nature.

While no therapeutic protein from transgenic microalgae has yet been commercialized, several studies have shown the real potential of this genetic approach [CAD 08, CAD 12, HEM 11].

The use of microalgae as "cell factories" requires the mastering of genetic modifications. It is currently difficult to make any comment on the number of microalgae, coming from different taxons, which have been genetically modified successfully. Generally, two paths of transgenesis have been used in microalgae: the nuclear path and the chloroplastic path. This last one is used in particular in the green algae *Chlamydomonas reinhardtii*.

Currently, around 20 therapeutic recombinant proteins have been produced in the chloroplast of the microalgae *C. reinhardtii*. The use of the chloroplastic path presents a considerable advantage to the nuclear path: it does not possess gene-silencing after gene modification. This process of gene silencing is often at the origin of a small accumulation of recombinant proteins via the nuclear path [POT 10], which explains why this path, unlike the chloroplastic path, is far less explored. The development of this production pathway is interesting for two reasons:

– being able to produce complex molecules requiring post-translational modifications;

– favoring secretion into the extracellular medium of proteins of interest, with the goal of minimizing the steps of extraction and purification of these proteins.

2.2.4. *Cosmetics*

Algae extracts are researched for use as emollients, incorporated into creams to prevent wrinkles, to stimulate the synthesis of collagens, without forgetting their part in the research on UV protection. While most of the assertions made in the advertisements may seem slightly exaggerated, the commercial perspectives largely justify the interest that is given to it. The same algae are seen again, *Spirulina* and *Chlorella*, in anti-aging products, regenerative products [SPO 06], or more originally, *Botryococcus braunii* [BUO 12].

2.2.5. *Industrial application*

The ability of some species of microalgae to accumulate molecules and physically elaborate structures in response to the biotopes that they colonize is a measurement of the diversity of the habitats previously mentioned.

2.2.5.1. *Silica and calcite*

Diatomit, extracted from primary sedimentary rocks, is largely used in filtration but can otherwise be used as an absorbent, a mortar, an abrasive material or as a complementary powder in horticulture. It has been shown that silica extracted from cultivated diatoms has an exploitable surface 80 times larger than that of diatoms of geological origin [CSO 99]. Silica is produced by phytoplankton in a lace structure with unmatched precision. The mastering, by genetic engineering, of the architecture of the fixing of silicon on the frustules of diatoms is thus envisaged to obtain new generation electronic components [BAO 07]. All these applications have been reviewed by [GOR 09].

2.2.5.2. *Emulsifiers*

Emulsifiers are components used to mix two immiscible substances. The classic example is of water and oil and the applications are common in the food industry and in cosmetics. Research in this domain is still quite limited. There is however an increasing association of microalgae with this idea [SCH 13].

2.2.5.3. *Depollution*

Microalgae are found to be associated with the treatment of urban effluents by wetpark. This technology, which remains fairly cheap, has been used for many years as a tertiary treatment system in the purification process. The improvement of the process, notably by the mixing of the lagoons, allows in certain cases an increase of both the polluting charge applied, initially smaller than in other systems, but also the efficiency of the purification. In these open basins, microalgae and bacteria work tightly together. By providing oxygen to the bacteria, the microalgae favor decomposition and the assimilation of organic substances. In return, the bacteria provide the microalgae with the CO_2 necessary for their metabolism. This synergy presents the advantage of allowing the elimination of toxic compounds, all the while limiting the risks of dissemination in the form of aerosols, favored notably by mechanical aeration systems. The oxygen produced by the microalgae favors a reduction of the sizeable energy requirement that its continuous supply entails, as in the systems of "activated sludges". Several pilots are being trialled according to this strategy [ZHA 13].

2.2.5.4. *Assimilation of nitrogen and phosphorus*

An important physiological characteristic of microalgae resides in a high intracellular concentration of proteins, and this is the case for most species.

Also, the metabolic activity necessary for the synthesis of these molecules requires an important intake of nitrogen. The synthesis of membranes (in the form of phospholipids) and of nucleic acids (DNA and RNA) calls upon phosphorus residing in the medium. This physiological activity associated with rapid growth makes microalgae big consumers of nitrogen and phosphorus. This characteristic can be observed notably by massive efflorescence in the natural environment, where anthropogenic output of these elements is great.

2.2.5.5. *Fixing heavy metals*

This aspect is one of the remarkable properties of these cells and is the subject of numerous studies and publications (33 publications in 2012 with the keywords "microalgae + heavy metal"). To meet their needs in vital metallic elements, microalgae use two main mechanisms. First of all, particular physico-chemical properties of their cell walls allow the trapping of metals to then move them into the cell, but also carry out the synthesis of important amounts of EPS. The hydrophilic and polyanionic compounds form a gangue, retain water and trap the cations, allowing the cell to resist desiccation.

These properties of accumulation have potential for applications in environmental engineering via the detoxification of environments polluted by heavy metals (lead, arsenic, mercury, cadmium, chrome, etc.) [SOU 12] or even in the salvaging of metals of interest such as gold [LUA 11], and could be applied for uranium or even copper. For a more general review of this domain, known "biosorption", see [GAD 09].

2.2.6. *Microalgae as fuel sources*

2.2.6.1. *Generalities*

As for other terrestrial biomasses, several paths of energetic valorization can be envisaged. The production of a raw algal biomass can be integrated into the more classical areas of energetic valorization, such as the organic matter made by terrestrial plants. These follow the processes of thermochemical conversion, liquefaction or even gasification. This biomass can thus be used as a substrate for the production of methane by anaerobic digestion, or for the production of ethanol by alcoholic fermentation. A second path is centered more around molecules synthesized which present high energetic potential: hydrogen and lipids.

The production of hydrogen by microalgae was observed for the first time at the end of the 19th Century. Since then it has remained the subject of intense research. The metabolic and molecular mechanisms involved are increasingly better understood, however, the problem of low productivity still needs to be solved, and makes this area difficult to exploit.

The idea of the use of microalgae to produce lipids for use in fuel is not a new one, even though the first proper advances only appeared in the 1970s, as illustrates the resulting report by the National Renewable Energy Laboratory (NREL) of the US energy department covering the years 1978–1996 [SHE 98]. Just like superior oleaginous plants, microalgae have the ability in certain conditions to accumulate an important amount of lipids. However, compared with these, they present *a priori* a large number of additional advantages, which contribute along with others to their recent "popularity", among which we will list:

– higher photosynthetic yield;

– no conflict with food supply;

– non-conflicting water management, if seawater culture;

– higher growth per Acre;

– high metabolic plasticity;

– mastering of the cycles of nitrogen and phosphorus;

– possible coupling with an industrial source of CO_2;

– possibility of continuous production;

– many valuable subproducts.

Several issues do arise, however, despite this list of advantages. Whatever the productivity obtained (even if, as we will see, it is very high) the surfaces of the cultures must be very large and the productions must be significant. In terms of mass energy, microalgae with energetic vocations will fare better in countries with large non-cultivatable spaces. A way of increasing productivity per unit of space consists in exploiting the undeniable advantage of being able to cultivate microalgae in large volumes. This option is followed by an increase of the price of a liter of oil, which is already high.

Recent reviews are mounting, discussing, analyzing and critiquing all of the supposed advantages of microalgae. For a list of the prospective publications see http://wwz.ifremer.fr/pba/ Presentation. For a list of the pertinent issues, we can look at "the Turquoise Book", a collective work by the French specialization currently in creation[2].

The main factors influencing[3] the production of lipids are linked to stress: low temperature, intense luminosity, low levels of nitrogen, severe phosphorus deficiency, lack of silica in the diatoms and alkalinity. Each of these parameters has a different effect, and analyses, calling upon factorial designs allow us to estimate the interactions between factors. Figures found in the literature range between 10 and 100 $g.m^{-2}.day^{-1}$ of dry biomass. For comparison's sake, cultures on the French Atlantic coast reach 10 $g.m^{-2}.day^{-1}$ (36.5 $T.ha^{-1}.year^{-1}$), a conservative figure, meaning that, for an alga containing 50% oil, the equivalent is of 18 tons of $oil.ha^{-1}.year^{-1}$. This number is far from the 100 tons of oil estimated by *Oilgae*, but the improvement margins, as much in the production of biomass as in the

2 www.adebiotech.org/home/img/algues/LIVRE_TURQUOISE-V.screen.pdf.
3 www.oilgae.com.

selection of super-producing strains, or even in the metabolic orientation during the different phases of culture, are important factors when starting with such a pessimistic calculation.

2.2.6.2. *Energetic yields*

One way of covering the energetic point of view is with a theoretical approach, such as the transfer of solar energy by photosynthesis. Experiments in the laboratory [BEN 97] have shown that 10 moles of photons (217 KJ.(mole photons)$^{-1}$) are required for the fixing of one mole of carbon (475 KJ.(mole C)$^{-1}$) by photosynthesis. The theoretical energetic limit is therefore 22%. Only 45% of the solar spectrum is exploitable by photosynthesis (Photosynthetically Active Radiation (PAR)), which decreases the maximum conversion efficiency of solar energy by photosynthesis to a value of 10%. In reality, the yields registered in experimental cultures are even lower, and it seems reasonable to retain an effective yield of around 3%. Moreover, we can consider that the lipids represent around 30% of the carbon fixed by photosynthesis, which leads to a photosynthetic yield for the synthesis of lipids of around 1%.

Based on this hypothesis, [CAD 08] presents potentials for the production of lipids in the order of 30–100 T.ha^{-1}.year^{-1} in Southern Europe. These figures must be compared with the 6 T.ha^{-1}.year^{-1} of the oil produced by palm trees.

2.3. The culture of microalgae

The use of microalgae in biotechnology increased in popularity in the 1990s [APT 99]. Today, most commercial production remains based on empirical knowledge. Information on the ecophysiological needs of species currently used or envisaged to be used in certain applications is rarely available, or patchy in scientific literature. Surprisingly, while they are fundamental for the optimization of productions, basic knowledge, such as the influence on growth or the biochemical composition of light, temperature, of the availability of dissolved carbon or nutritional requirement, is often yet to be established. In the same manner, despite a lot of research and development around culture systems [OGB 03], progress still needs to be made to improve the economy of phytoplankton production.

2.3.1. *Ecophysiological needs*

2.3.1.1. *Light*

Being autotrophic organisms, microalgae are able to conduct photosynthesis. This process allows the conversion of radiation energy, in this case light (Figure 2.3), into exploitable carbohydrates by the cell. Only 2–3% of the incident energy is actually used by the vegetal biosynthesis pathways [MEL 09].

Figure 2.3. *Toric photobioreactor, with a geometry particularly adapted to studies of the effects of light on the growth of microalgae (photo Kaas/Ifremer)*

This biosynthesis is carried out in two distinct phases.

In the first phase, the light or "photochemical" phase, photons are absorbed by the chlorophyll, which goes into an activated state and only goes back to its initial state by releasing the captivated energy: this energy is used to carry out the photolysis of water which then releases H^+ ions, oxygen and electrons. Two metabolic pathways redistribute the potential energy of the photons and of the electrons through a series of transporters located in the membrane of the chloroplast. One produces energy directly available for use by the cell in the form of adenosine triphosphate (ATP) and the other produces reducing organic molecules [FAR 06].

In the second phase, the dark phase, the reduced organic molecules, and part of the ATP formed during the light phase, are used in the synthesis of phosphorylated sugars (pentose pathway). Variations of this photosynthetic cycle exist. In particular we can cite the C4 cycle of tropical poaceae and the cycle of succulent plants. The biosynthetic pathways of certain bacteria are also different. These mechanisms linked to CO_2 have been identified in certain microalgae but their exact role and their level of activity remains contested [HAI 13].

The central molecule, and precursor of the reactions of photosynthesis, is chlorophyll, a tetrapyrrolic molecule with an atom of magnesium at its core. This is found associated with proteins included in lamellar structures, called thyllacoids. These lamellar structures are organized into super-structures called *granum*, which are most often gathered together into chloroplasts [FAR 06].

In the natural environment, microalgae live in the euphotic zone of water masses [KIR 94]. The ability of microalgae to become acclimatized and to adapt to this zone has allowed them to spread out according to their sunlight requirements so as to obtain the best conditions to carry out photosynthesis. This depends directly on the incident luminous flux on the cell, whose power per unit of surface is expressed in $W.m^{-2}$ (equivalent to $J.s^{-1}.m^{-2}$). This flux is referred to as "irradiance" and is expressed in $\mu mol.photons.m^{-2}.s^{-1}$. Without the intervention of another phenomenon that would limit the physiology of the microalgae, irradiance directly determines its growth rate (μ) which, in function of the irradiance (I) (Figure 2.4), follows a typical curve, whose optimum is dependent on the species. The growth rate increases rapidly for low irradiances, lower than a few hundred $\mu mol.photons.m^{-2}.s^{-1}$, then reaches an asymptotic optimum, and finally decreases as soon as the process of photoinhibition intervenes, as well as the oxidative stress that results from it.

Light is also defined by its spectrum, which is altered when it penetrates water. In solar electromagnetic radiation, visible light lies between 400 and 70 nm, and represents 40% of the incident energy. Due to the pigments that make up the Light Harvesting Complex (LHC) of the microalgae's photosystems, it is within this domain of wavelengths that photosynthetic activity takes place. It defines the PAR. This is not to be confused with the Photosynthetically Usable Radiation (PUR), which groups together all of the wavelengths actually usable by the microalgae and which depends on the pigment composition of a species in a given state. This composition determines the absorption efficiency of the photonic energy and of the electronic transfer of LHC as well as the aptitude of the LHC to dissipate excess illumination. The acclimation abilities of the microalgae, with regards to the quality and the quantity of light, are linked to the plasticity of the biosynthesis of pigment. This acclimation is more guided toward the optimization of photosynthetic activity than toward its maximization [MAC 02].

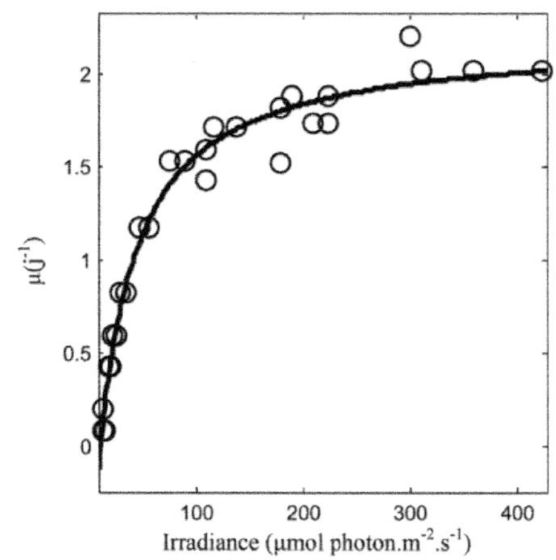

Figure 2.4. *Variation of the growth rate (μ) of Tisochrysis lutea in function of the irradiance (data PBA-Ifremer)*

Access to an optimal quantity of light for each of the cells of the microalgae is a classical problem met by the production systems of microalgae, as they need to obtain high levels of productivity, high biomass, and this within sizeable volumes.

The more the concentration of a culture increases, the more the phenomenon of luminous attenuation accentuates. This is often represented by the optical law of Beer-Lambert: the self-shading of the cells generates a light gradient, with a rapid decrease of the irradiance in function of cellular depth and concentration [LEE 99]. In dark or thick cultures, this results in illuminated zones where photosynthesis dominates, and dark zones where respiration is the main phenomenon. The light gradient can therefore lead to significant losses of productivity.

In an attempt, to increase the access to light of all the cells of a culture, the decrease of the thickness of the culture is a strategy to optimize the penetration of light, although this does cause a decrease of the volume of the culture for a given surface. Another strategy involves the agitation of cultures. The various agitation devices available each have pros and cons, in terms of mixing efficiency, biological constraints and exploitation costs. The management of light is a key factor in the growth yields, notably in the case of controlled production areas like photobioreactors (PBRs). This issue has experienced increased interest since the 1990s because of the use of customized light due to light-emitting diodes (LED). These controllable lighting devices offer possibilities of biochemical orientation in controlled algal productions, whether during growth or in the maturation process of the cells. Up until now, the studies conducted have involved the metabolic effects of the wavelengths of lighting [MAR 13], of high values of irradiance (several thousand $\mu mol.photons.m^{-2}.s^{-1}$) and irradiance frequencies [YEH 09].

Research and development in the domain of light remains extensive: which spectrum to use to orient biology? How to manage mixes of wavelengths? What are the limits of the scale transfer of LEDs? The technologies of light sources continue to evolve. The use of organic LEDs or polymer LEDs is promising, as is the use of flexible materials to take on the shape of the culture receptacles so as to reduce light loss.

In the case of external productions, where the light is highly available in according to the latitudes, the natural phenomena of night/day cycles and of the climate are dynamic parameters that constantly modify the incident light. While these variations limit the access of outside cultures to the highest levels of productivity, the natural plasticity of the cells, the polymorphism and the genetic potential of a population of microalgae ensure a certain ability of acclimatization and of resilience. Outside, developments in light can involve the modification of the solar spectrum so as to control metabolism. The use of films or plastic greenhouses that filter some wavelengths is one example. Such setups, whose costs do need to be evaluated, can be a way of equally dealing with rainy episodes, evaporation and could be a potential way of helping with a better thermal management of cultures.

2.3.1.2. *Temperature*

Whether the microalgae are cultivated outside or in controlled conditions, knowledge of each species' reaction to the temperature is primordial. In the case of outside productions, it allows for the choice of species that are particularly adapted or for a succession of species that will accommodate to seasonal changes in conditions. In controlled conditions, the leading of cultures to their optimal temperature leads to an optimization of the productions.

Generally, microalgae that present a high temperature optimum have high growth rates. The sensitivity of rate of growth to temperature is characterized by strong asymmetry: below the optimum, a drop of temperature leads to a gradual decrease of the growth rate. Contrarily, temperatures above the optimum affect growth in an abrupt manner [EPP 72] (Figure 2.5).

Beside the growth rate, temperature can affect other characteristics of the biomass and notably its composition. It is well known that the temperature of the culture influences the amount of pigment it contains [ZUC 01], notably of chlorophyll, with the appearance of phenomena of chlorosis (destruction of the chlorophyll) at low temperatures [GEI 87]. Lipid content and composition are also sensitive to temperature [CON 09], resulting in an increase of the lipid concentration and of the degree of unsaturation of the fatty acids for low temperatures. [ZHU 97] shows clearly, in *Isochrysis galbana*, the influence of a drop in temperature from 30°C to 15°C on the increase of the concentrations observed a 18:3(n-3) and 22:6(n-3). These two fatty acids depend on an omega-3 desaturase, whose expression is controlled by the temperature.

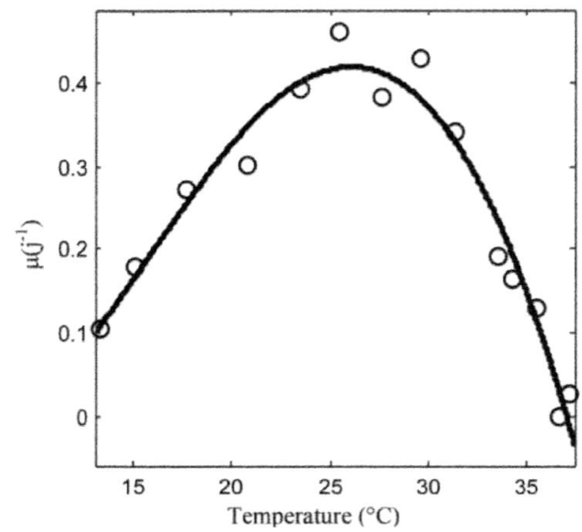

Figure 2.5. *Variation of the growth rate (μ) of Dunaliella salina in function of the temperature (data from PBA-Ifremer)*

2.3.1.3. *pH and inorganic carbon*

CO_2 is the main anthropogenic greenhouse gas. The capture of CO_2, coming mainly from thermal power stations, metal and cement production, is a major concern for those involved in the industry.

As opposed to terrestrial plants, which have direct access to atmospheric CO_2, microalgae must use dissolved inorganic carbon (DIC) in the aqueous medium, in the form of bicarbonate (HCO_{3-}) or CO_2. Especially in the marine environment, all other things being equal, pH is closely linked to the amount of inorganic carbon dissolved and governs the balance between the different forms of carbon [DIC 87]:

$$CO_2(gas) + H_2O \Leftrightarrow CO_2(dissolved) + H_2O \Leftrightarrow H_2CO_3 \Leftrightarrow HCO_3^- + H^+ \Leftrightarrow CO_3^{2-} + 2H^+$$

In surface marine waters, the pH is generally 8.2. The dissolution of atmospheric carbon in the medium is the limiting step of this system. The fixing of carbon by the photosynthetic activity of the algal biomass can reduce the availability of the DIC at levels that mean that primary productivity is limited, resulting in a decrease of the growth rates and in accumulation of carbon reserves. The photosynthetic activity of microalgae therefore results in a significant increase in the pH of the cultures, which can reach values greater than 10 at the end of the culture. Microalgae have therefore developed active carbon transport systems and specific enzymes (carbon anhydrases) that are released into the external medium, allowing them to efficiently access low ambient concentrations of carbon [BHA 02]. The sensitivity of microalgae to the pH of the medium is seen in a curve of the response of the growth rate in the shape of symmetrical convex bell (Figure 2.6). Values of pH that are more acidic than the optimum can lead to the excretion of part of the carbon fixed by the algae, while higher values of pH limit the fixing of the carbon. There are very few studies available on the influence of pH on the growth of microalgae. It would seem that for marine algae, the optimal pH is often located around the pH of seawater. However, it has been shown that *T. lutea* (*Isochrysis affinis galbana*, clone T-Iso) has an optimal pH located around 7, a value which is significantly different from that of seawater [MAR 12].

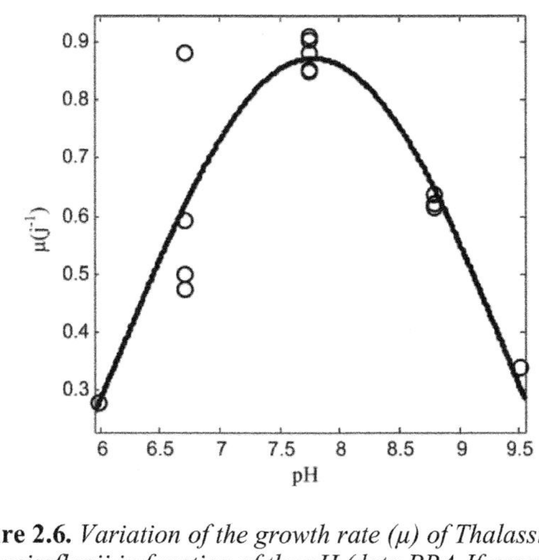

Figure 2.6. *Variation of the growth rate (μ) of Thalassiosira weissflogii in function of the pH (data PBA-Ifremer)*

From the phytoplankton productions point of view, these physico-chemical aspects have important repercussions on productivity. For this reason, it is advisable to regulate the pH of the cultures so as to ensure a constant availability of DIC. This regulation is classically carried out by injection of gaseous CO_2 or of HCO_3^- into the cultures.

It takes between 1.5 and 2 kg of CO_2 to produce 1 kg of algal dry weight. Here again, the distinction must be made between the yields provided by open culture processes and production carried out in PBRs. A prospective analysis carried out as part of the Shamash research project (a project supported by the French National Research Agency) found a possible fixing between 13 and 280 tons of CO_2 per hectare and per year in an open system.

2.3.1.4. *Nutrition*

Microalgae are generally considered to be photoautotrophic organisms, meaning that they are able to develop from light energy and from nutrients. However, some species are also able to develop from the chemical energy contained in organic carbon. They are thus heterotrophic for carbon.

2.3.1.4.1. Mineral nutrition

When they are "operating" in a photoautotrophic regime, microalgae take all the required mineral nutrients from their dissolved form in the medium. We can make the distinction between macronutrients and micronutrients depending on whether the quantitative needs of the microalgae are high or low. For the growth of marine microalgae, the essential macronutrients are carbon, nitrogen and phosphorus. [RED 34] observed that these three elements are included in the elementary composition of phytoplankton in uniform proportions, statistically, and with a molar ratio of 106:16:1, called the Redfield ratio. The micronutrients include elements (iron, cobalt, zinc, manganese, magnesium, etc.) that are vital for certain key enzyme activities as well as other intracellular functions.

Besides carbon, which makes up the carbon backbone of molecules, cellular nitrogen is found majorly and in equal proportion (around 30%) in the nucleic acids (DNA and RNA) and proteins.

Chlorophylls represent another significant stock, around 10%, of intracellular nitrogen. In the same manner, a large part of intracellular phosphorus, around 40%, is involved in the composition of the two types of nucleic acid. Membrane phospholipids represent up to 20% of cellular phosphorus [GEI 02]. Finally, while they only represent a small proportion of cellular phosphorus, the adenosine phosphates (ATP, ADP, AMP) and NADPH play a primordial role for the cell's energy and reductive powers.

As mentioned previously, carbon is absorbed by microalgae in the form of CO_2 or dissolved HCO_3^-. Mineral phosphorus is bioavailable in the form of phosphate (PO_4^{3-}). However, in the case of a strong limitation for the phosphate, microalgae can access organic phosphorus through the use of extracellular phosphatases [KUE 65]. The forms of mineral nitrogen accessible by microalgae are more diverse and comprise nitrates (NO^{3-}), nitrites (NO^{2-}) and ammonium (NH_4^+). The most reduced forms are more readily assimilated due to a smaller energy requirement. On the other hand, at high pH levels, the transformation of ammonium into ammoniac can generate high toxicity. In conditions of limitation by mineral nitrogen, microalgae are, like for phosphorus, able to absorb different forms of organic nitrogen: the use of urea and free amino acids in this way is well known [SOL 10].

Whichever the element concerned, microalgae possess very efficient absorption systems that allow them to take elements from the medium, down to concentrations lower than a micromole per liter [FAL 75]. At these low concentrations, the transport systems involved are said to be active and as a result their operation requires energy provided by photosynthetic activity. While nitrogen transport systems are relatively well-studied, with a diversity of transporters, more or less specific for one chemical species, and variable affinities and capabilities for their substrate [REX 02], studies on phosphate transporters remain uncommon.

Under the pressure of a limitation, microalgae are able to accumulate the nutrients present in excess in the medium, well beyond their needs, resulting in a phenomenon called "luxury consumption". Excess nitrogen can thus be stocked in the form of nitrates or ammonium depending on the species [LOM 00], while reserves of phosphorus are held in the form of polyphosphate granules [EIX 05]. For example, in *Selenastrum minutum*, [ELR 85] showed maximal stocking capabilities in excess of 16 times the requirements for growth, while for nitrogen, this capability is four times smaller.

The availability of nutrients conditions and limits the development of microalgae populations. Depending on whether the limiting factor modulates the growth rate or the maximal level of the population, this is either a Blackman limiting factor [BLA 05] or a Liebig limiting factor [VON 55]. To understand the impact of nutritional limiting factors, Droop [DRO 68] proposed an increasing and hyperbolic relation between growth and cellular content of the limiting element (notion of cellular quota). This relation has been used successfully to describe microalgal growth as part of the concept of the law of the minimum [RHE 78], which postulates that the element that governs growth is governed by the element with the shortest supply compared to algae need. More recently, the concept of co-limitation has partially questioned the paradigm of the law of the minimum [SAI 08]: in some situations, several nutritional elements, whether biochemically linked or not, can work together to limit growth at the same time. Varied examples of co-limitation involve substrates and co-enzymes of a particular enzyme path, like carbon and zinc in carbonic anhydrase [BUI 03]. Similar co-limitation phenomena were recently also assumed between nitrogen and phosphorus [BOU 10, PAH 09].

The notion of limitation by the nutrients has important implications for application. Nutritional limitation, linked to the strong metabolic plasticity of microalgae, can result in the accumulation of stocked carbon (sugars and lipids). This observation is largely used as part of research on the production by microalgae of energetic lipids for use in biodiesel [CAD 08].

This ability to use solar energy is by far the most used. The large global microalgae productions mainly call upon the photoautotrophic pathway.

The company Mera Pharmaceuticals, located in Hawaii, produces a biomass of 6.6 $T.year^{-1}$ due to use of a closed tubular PBR. Similar systems are also used by the company Algatechnologies, in Israel, and Fuji Health Science, in Hawaii. However, due to the low rate of growth of *Haematococcus* and the production protocol that requires two steps, the astaxanthine produced using these methods is hardly competitive with synthetic pigment [GUE 03].

The production of ß-carotene by *Dunaliella* sp. is also very important in the world. It represents a level of photoautotrophic production of 1.2 $T.year^{-1}$ globally. Two types of production coexist. The companies Betaten and Aquacaroten, located in Australia, cultivate this microalga in lagoons without agitation. Betaten lists annual productions of around 13 tons of ß-carotene for a culture surface of around 400 hectares. The associate production costs appear relatively low due to a favorable climate and the absence of energy consumption linked to more intense culture systems, like *raceway*. Several studies describe *Dunaliella* trials in closed PBRs. These trials, however, have not yet been conducted on an industrial scale [DEL 07].

Many small companies produce a variety of microalgae, cultured photoautotrophically for their high contents of polyunsaturated fatty acids like DHA and EPA. This is notably the case for the company Innovative Aquaculture Products Ltd., established in Canada, which produces the haptophyte *T. lutea* or BlueBiotech ln T in Germany or even Innovalg in France, which produces the diatom *O. aurita*, co-cultured with the microalga *Chondrus crispus*, intensively in *raceways* with surfaces of up to 1000 m^2.

Finally, one of the biggest PBRs for the production of microalgae photoautotrophically is established in Hanover, Germany. With a total volume of 700 m^3, this tubular PBR, developed by [PUL 01] and recently acquired by the company Roquette, is capable of producing between 130 and 150 tons of dry microalgal matter per year.

2.3.1.4.2. Organic nutrition

Heterotrophy is defined as being the ability of an organism to develop on the basis of organic carbon produced by another organism. A medium for heterotrophic culture is thus made up of the same mineral elements as those necessary for photoautotrophy, but also provides a source of organic carbon.

Work on the heterotrophic pathway in microalgae was started as early as the 1960s and showed that some species were able to develop on the basis of organic carbon substrates. Among all of the known species, only a small number demonstrated this ability [PER 11], and it would also seem that commonly cultivated marine microalgae present relatively undeveloped abilities for heterotrophy [UKE 76]. Since the 2000s, the heterotrophic pathway is increasingly researched activity, with a lot of work from Chinese teams [LIA 09] as part of the development of the production of algae oils.

The main advantages of heterotrophy relate to:

– potential productions in volume, in any kind of fermenter and no longer only on the surface like in the autotrophic systems;

– the reduction of production and maintenance costs [PER 11].

The advantage of this trophic pathway for industrial cultures aiming at products with little added value – such as biofuels – must however be relativized because of the disadvantages it presents compared to the more classic autotrophic culture:

– the sizeable volumes involved in "biofuel" applications would require access to a cheaper resource of organic carbon;

– the respiratory activity of cells, linked to the degradation of organic carbon sources, generates CO_2, which can be remedied in the autotrophic pathway;

– the heterotrophic path requires axenic conditions or at least very low contamination levels to avoid competition with bacterial communities that usually have higher growth rates than the microalgae. As underlined by [BUM 11], even the smallest microbial contamination introduced after inoculation can easily overtake the microalgal biomass in the competition for organic carbon. The maintenance of axenic conditions during the production period thus requires precautions of culture protocol and the use of particular materials. Heterotrophic productions should therefore be reserved for high-value markets.

Generally, species capable of heterotrophy show growth performances and maximal biomasses that are greatly superior to those recorded in photoautotrophy [LEE 97], by using a variety of organic substrates (glucose, glycerol, acetate, malate, lactate, glutamate and ethanol) depending on the metabolic capabilities of the species involved [BUM 11]. For example, high concentrations of biomass (45 $g.L^{-1}$) and volume productivities of 20 $g.L^{-1}.day^{-1}$ have been obtained due to heterotrophic *Nitzschia alba* cultures [GLA 94]. The direct consequences are lower costs in production, maintenance and harvesting. Production by heterotrophy in a fermenter is well mastered in Japan and Korea, mainly for aquacole and nutraceutical applications [LEE 97]. The company Martek, in the USA, produces DHA with the help of heterotrophic *C. cohnii* cultures [MEN 09].

Quite surprisingly, some pigments are produced by microalgae, even in the absence of light [PER 11]. Thus, [SHI 00] obtained lutein contents between 68.42 and 83.81 $mg.L^{-1}$ in the heterotrophic *Chlorella protothecoides* cultures. [WAN 08] also carried out production of astaxanthin with *Chlorella zofingiensis* cultured on glucose. This study even suggests that the *maxima* of biomass and of astaxanthin production can be obtained simultaneously with the help of a single-step culture, instead of the two-step culture usually carried out for *Haematococcus sp.*

The mixotrophic pathway is an intermediary between strict heterotrophy and autotrophy; it allows the algae to benefit from both organic carbon and CO_2, through photosynthesis [HER 11]. It appears as a more promising alternative to autotrophy than heterotrophy, in as much as it combines the advantages of each of the previous pathways, leading to synergic effects: the growth rates recorded for the species concerned are indeed greater than those from the heterotrophic pathway [SCA 10]. [YAN 00] obtained biomass yields, calculated on the basis of the energy provided, four times greater for *Chlorella pyrenoidosa* cultures produced in mixotrophy, compared to photoautotrophic cultures. They also showed that alternations between photoautotrophic phases and heterotrophic phases could lead to even higher yields than for true mixotrophy. [PUL 04] observed that the growth rate in mixotrophic cultures of *Chlorella vulgaris* and *Haematococcus pluvialis* was the sum of the growth rates obtained in strict heterotrophy and strict photoautotrophy. Mixotrophy can therefore allow the overcoming of limitation caused by the availability of light in high-density cultures. [COM 94] underline that this mechanism is important in the case of *Scenedesmus obliquus* and it seems to be shared by many mixotrophic species.

Mixotrophy also provides an effective way of increasing the productivity of light-inducible metabolites. [STA 98] also obtained in *Galdieria partita* contents of chlorophylla, carotenoids, phycocyanin and allophycocyanin that were greater in mixotrophy in comparison to the results obtained in heterotrophy. Analogous results are reported by [SHI 99] with lutein and chlorophyll contents in *C. protothecoids* that are greater in mixotrophy.

Finally, mixotrophic activity, combining both respiration and photosynthesis, allows the consideration of neutral or positive carbon balances. However, as for heterotrophic cultures, the controlling competition with bacterial communities remains unstudied.

To a certain extent, microalgae cultures become independent, or at least only slightly dependent, on any source of light. This characteristic contributes to a smaller occupation of space, a decrease in the investment linked to production structures, a production rhythm independent of the nycthemeral cycle and, to a lesser extent, of the seasons. Moreover, the cultivation of microalgae on organic carbon substrates allows us to consider the purging of sources not only rich in organic carbon but also in nitrogen and phosphorus.

2.3.2. *Productions and productivities*

2.3.2.1. *Modes of culture*

There are three different modes of production: the discontinuous mode, the semi-continuous mode and the continuous mode of cultures. In the case of discontinuous cultures, the various nutritional elements are only distributed at the start of the production, and the harvesting is carried out at the end of the cultivation period. Semi-continuous cultures are partially harvested at regular intervals, and also resupplied in the same manner with nutritive medium. Continuous cultures allow for the constant harvesting of cells, as the supplying of the medium nutritionally is ensured continuously. This method of production allows for stable culture parameters throughout the entirety of the length of production, and therefore allows for the production of a biomass of constant quality. It also reduces production costs by minimizing the amount of human work.

2.3.2.1.1. The discontinuous mode [BAI 86]

If the PBR or the basin is only supplied with nutrients once, without any further supply, the system functions discontinuously, called "batch mode". The inoculated cells experience a period of adaptation to the new environmental conditions, of various lengths, called latency phase, and then develop and express all of their potential when the amounts of nutritional salts and light are not yet limiting factors. During this period, called the exponential growth phase, the algae reach their highest specific growth rate in function of the conditions imposed on them. If we call x the concentration of microalgae in the expressed culture volume (either in $g.L^{-1}$ or in the number of cells per liter), we can define the growth rate of the culture as being the increase of the cellular concentration per unit of time [2.1]:

$$V_{vol} = \frac{\partial x}{\partial t} \text{ expressed in } g.L^{-1}.h^{-1} \text{ or in number of cells } m.L^{-1}.h^{-1} \quad [2.1]$$

By dividing the growth rate by the amount of algae present at time t, we define the specific growth rate (μ), independent from the volume of the culture, which is an essential characteristic of cells in culture [2.2]:

$$\mu = \frac{V_{vol}}{x} = \frac{\partial x}{x \times \partial t} \text{ expressed in h}^{-1} \text{ or in days}^{-1} \quad [2.2]$$

The specific growth rate expresses the ability, for example, of a gram of algae, to produce μ grams of algae per unit of time (in hours or in days), or the ability of cell to produce μ cells per unit of time.

It comes naturally that:

$$\mu \times \partial t = \frac{\partial x}{x} \quad [2.3]$$

which we can integrate between t_0 and t_1 when the algal concentration passes from x_0 to x_1 [2.4]:

$$\mu(t_1 - t_0) = \ln(x_1) - \ln(x_0) \quad [2.4]$$

$$\mu = \frac{(\ln(x_1) - \ln(x_0))}{(t_1 - t_0)} \text{ expressed in h}^{-1} \text{ or in days}^{-1} \quad [2.5]$$

The generation time *(t_g)* is another value that characterizes the abilities of these organisms in a determined situation. It represents the time taken by a population to double, either in weight, or in numbers. By going back to equation [2.4], we can replace x_1 by $2x_0$ and we obtain:

$$\mu = \frac{(\ln(2 \times x_0) - \ln(x_0))}{(t_2 - t_0)} = \frac{\ln(2)}{(t_2 - t_0)} \quad [2.6]$$

$(t_2 - t_0) = \frac{\ln(2)}{\mu} = G$ is the generation time expressed in hours or in days [2.7]

2.3.2.1.2. The continuous mode

If the bioreactor is constantly supplied with medium, the mode of culture is said to be continuous. For the volume present at any moment within the system to remain fixed, the charge introduced into the reactor is equivalent to volume removed. The continuous mode can be started either from a culture in discontinuous mode, or directly [BOU 03]. In function of the imposed culture conditions, the population of the cells will reach a level of equilibrium that is a function of the amount of light dispensed, of the nutrient content of the renewal rate imposed by pump bringing the culture medium. This state is not very stable, and any modification, even minimal, of the conditions governing the reactor will lead to a change in this equilibrium level. At equilibrium, the amount of algae produced per unit of time is equal to the amount exported by the flow of the pump [SCR 99]:

$\mu \times x \times V = Q \times x$ with V = reactor volume (L) and Q = flow (L.h^{-1}) [2.8]

$$\mu = \frac{Q}{V} = D \text{ corresponds to the dilution rate of the reactor} \quad [2.9]$$

By imposing a determined flow, we are fixing the specific growth rate of the microalgae at equilibrium. If the dilution rate overtakes the maximal specific growth rate of the species in the conditions of the culture, the cells will be washed out and the cellular concentration will irremediably decrease until there are no more cells in the culture volume [BAI 86].

It is possible to lead cultures continuously according to different principles which, at equilibrium, provide the same results:

– the chemostat principle: in this operating mode, the culture is continuously supplied with nutritional medium with a pump whose flow rate is fixed at a certain value. By fixing the flow, the specific growth rate is imposed and the culture will stabilized at equilibrium around a cellular concentration undetermined at the start;

– the turbidostat principle: the objective of this type of culture is to reach a determined cellular concentration by using the flow of the pump at a continuous measurement of the biomass or of the cellular concentration present in the reactor. This requires the use of a measurement system allowing us to consider the charge present in the reactor. Often, optical measurement systems are used to link the cellular concentration to light absorption. By fixing an optical instruction, the optical system will dilute the culture when this value is overcome. After a certain amount of time has passed, and in function the hysteresis of the measurements, the pump will be controlled near-continuously according to a flow rate corresponding to a dilution rate equal to the specific growth rate of the cultivated algae.

2.3.2.1.3. The discontinuous supply mode, or *Fed Batch*

The discontinuous supplied mode is characterized by the addition, in a predetermined or controlled manner, of nutrients in the PBR at certain periods. The addition of nutrients can thus be adapted to cellular concentration or the biomass present in the production area [BAI 86]. At the start of the culture, the reactor is only partially full, the new medium is brought gradually along with the development of the cells and the culture is usually finished when the reactor is full. This process is often used for bacterial or yeast productions as it allows the reaching of important levels of productivity. In the case of microalgae, the requirement of bringing light to allow the cells to grow significantly reduces the attraction of this type of production. Heterotrophic cultures can, however, benefit from this, and important levels of productivity can be reached. [BUM 11] report cellular concentrations in the order of 100 $g.L^{-1}$ for *Chlorella sp.*, *C. cohnii* or *Galdieria s.* cultured heterotrophically.

In this case, the amount of algae produced per unit of time is linked both to the variation of the volume of the culture and to the concentration in algae:

$$\mu \times V = \frac{\partial (V \times x)}{\partial t} \quad [2.10]$$

$$\mu \times \partial t = \frac{\partial (V \times x)}{V} \quad [2.11]$$

By integrating between time t_0 and t_1 and the volume V_0 and V_1, it results that:

$$\int_{t_0}^{t_1} \mu \times \partial t = \int_{V_0}^{V_1} \frac{\partial (V \times x)}{V} \qquad [2.12]$$

$$\mu \times t = \ln\left(\frac{x \times V}{x_0 \times V_0}\right) \qquad [2.13]$$

and the concentration at any moment within the reactor is an exponential function of time:

$$x \times V = x_0 \times V_0 \times e^{\mu \times t} \text{ expressed in number of cells} \qquad [2.14]$$

2.3.2.2. *The production systems*

Currently, the techniques for the production of microalgae are split between large-scale cultures in an open medium, lagoon, continuous flow system (*raceway* or circular basin) and closed systems, PBRs, plastic bags, sheaths, pipes (Figure 2.7). The choice between the different culture techniques is based on economic considerations, very much in favor of open systems, but which present risks linked to possible contamination and little amounts of control to counterbalance climate variations.

Today, most of the production of microalgae is carried out in external basins [SPO 06].

The distinction must be made between production in controlled PBRs with artificial or natural lighting, which allows for very high levels of productivity, with a circulating charge able to reach 4 $g.L^{-1}$, for a high cost in terms of both investment and operation (the reactors also allow to proceed with highly controlled productions for biotechnology and pharmaceuticals), from production in natural or tubular lagoons extensively with algal concentrations generally comprised between 0.4–1 $g.L^{-1}$ and for which the yields and the costs are lower.

Figure 2.7. *Photobioreactor JSP-120 for the automated production of microalgae in marine mollusk hatcheries (photo Kaas/Ifremer)*

2.3.2.2.1. Lagoons

One of the most important lagoon-based commercial productions of microalgae is carried out in Australia, where hundreds of hectares are used to cultivate *Dunaliella salina*. With a depth of 20–30 cm, they are only agitated by the wind and movements of convection. This extensive production is made possible by extreme conditions, notably very high salinity levels, which suit the growth of *Dunaliella salina* and limit the development of contaminants. This culture system also requires little land costs, due to the large surfaces required as well as the optimal climate conditions throughout the year. Moreover, seeing as productivity never exceeds 1 $g.m^{-2}.day^{-1}$, an efficient harvesting system is required [RIC 08].

2.3.2.2.2. *Raceways*

Raceways are currently the most commonly used production systems used for the large-scale cultivation of microalgae. The choice of this system of culture is usually based on economics, as the investment and operating costs are not high. They are usually cement or dug basins, covered in plastic films with a water height that does not go over 40–50 cm, to ensure light access, and not below 20 cm to reduce temperature fluctuation problems. Impellers ensure the agitation to maintain the algae in suspension. Compared to closed systems, the levels of productivity reached are not very high. The difficulty comes partly from the low effectiveness of agitation, which does not allow access to the light and to optimal gas–liquid transfers, and partly due to the lack of control over the ambient conditions. The cultures are submitted to the randomness of the weather. For example, in the case of strong rain, the salinity of the culture is reduced, the nutrients and the biomass are diluted, reducing productivity, without forgetting an important risk of contamination by other microalgae, protozoa and bacteria.

Geographical positioning plays a vital role in the productivity, which is estimated in a *raceway* to be between 10 and 25 $g.m^{-2}.day^{-1}$ [ROS 12], while the theoretical calculations would leave hope up to 60–198 $g.m^{-2}.day^{-1}$ [COO 11]. Average productivity levels observed for the diatom *O. aurita* by the Innovalg society in the "Pays de Loire" are of 10 $g.m^{-2}.day^{-1}$.

2.3.2.2.3. Photobioreactors (PBR)

PBRs are closed systems (Figure 2.8), lit up by a light source that can be natural when the PBR are placed outside [GRO 03] or artificial (fluorescent tubes, LEDs).

They can be equipped with different probes so as to control the kinetics of the growth of the microalgae (temperature, pH, agitation speed, renewal rate and lighting). Many different types have been developed. Some correspond to specific study requirements, in particular in the domain of ecophysiology, while others have been established with the goal of optimizing the commercial production of biomass.

The PBRs allow us to work in condition of asepsis, ensuring the lack of contaminations such as the predators of microalgae, undesired opportunist species, and to limit the bacterial load. If they are built from glass and stainless steel, they can be sterilized with high temperature steam. This type of material is very costly, and as a result most PBR are made of synthetic matter such as PMMA (poly(methyl methacrylate)). Sterilization is then chemical (bleach or peroxyacetic acid).

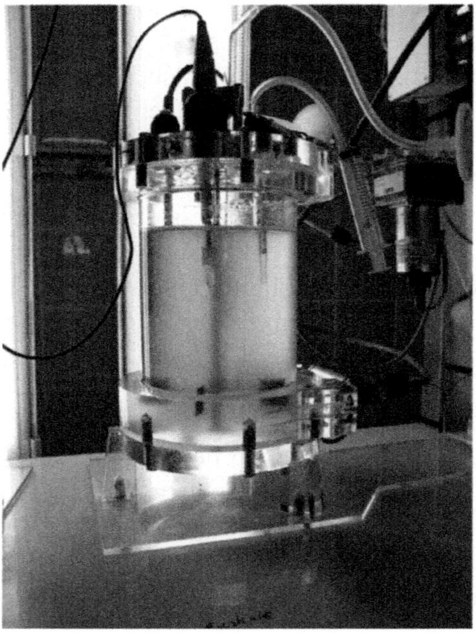

Figure 2.8. *Photobioreactor with instrumented turbine, used in ecophysiological studies in controlled conditions (photo Kaas/Ifremer)*

As part of studies on the effect of lighting, as much in terms of quantity as of the nature of the spectrum, the use of planar PBRs of reduced thickness is favored. This arrangement allows a homogeneous distribution of luminous energy inside the PBR and provides a representative measurement of the light that really penetrates the area.

This type of PBR, equipped with an absorption probe that activates a sampling pump, provides the possibility of working in turbidostat mode. This allows the maintenance of a biomass concentration that is constant and low, no matter the quantity and the quality of the light, as the growth rate indirectly steers the operating frequency of the sampling pump. This type of arrangement means that, due to the fact that the access to light is optimized, high levels of productivity can be achieved. However, these PBRs do not allow a volume increase. To conserve the benefit of the small thickness, and therefore of the access to light, but to increase the volume, either an increase in height or an increase of the length of the PBR is required. For a planar PBR of 1,000 liters and with a thickness of 6 cm of the curve number, 1.5 m of height and 11 m of length would be needed. Moreover, the levels of productivity then achieved would be reached at the expense of a large amount of water, the elimination of which would add a significant cost.

To deal with this problem of space, the choice of cylindrical PBRs is required [TSY 01]. This type of reactor can be made up of one or several tubes.

Column-type cylindrical PBRs are largely used to produce microalgae in mollusk hatcheries. To have a sufficient culture volume to deal with the needs of the larvae, these PBR have a capacity of around 300 l and a diameter in the order of 60–90 cm. The light parameter quickly becomes a limiting factor, as the cells of the microalgae limit the penetration of the photons in the PBR. Ring-type cylindrical PBRs partially overcome this problem of light access to the culture. They are made up of two concentric tubes of equal height. The lighting system is positioned in the central tube. The thickness of the water layers exposed to the lighting is less important and the biomass yield is therefore better.

Tubular cylindrical PBRs lead to a significant reduction of the light's access path. They can be made up of a long tube with a small diameter (between 1 and 5 cm), deployed in a serpentine form on the ground. The congestion of occupied surface is very high for a PBR of several hundred liters. They are therefore mostly reserved for outside use [TRE 98]. The other way to accommodate the long tube is to roll it around a support of up to several meters of diameter and of height [TRE 98]. The space occupied on the ground is then greatly reduced.

These two type of tubular reactor are dedicated in priority to outside installations.

The last type of tubular PBR is more suited for inside production. They are made up of vertical tubes of 50 cm to 1.5 m, linked together by flanges so as to form a closed loop [OLI 07]. These tubes, with a diameter in the order of 6 cm, can be deployed in a space on the ground and in areas of reduced height. For example, a PBR of 120 l containing the microalgae *T. lutea*, allows a productivity of 20–30 $g.m^{-2}.day^{-1}$ of dry matter in the laboratory. The agitation of the culture inside the PBR can be ensured by an air injection device, or *airlift*, by a stirring device or by a pump. Depending on the fragility of the species cultivated, it is important to take into account the shearing forces induced by the agitation system and by the geometry of the PBR. In as much as it constitutes a closed system, it is also necessary for the PBR to be equipped with a degassing system of the oxygen produced (in the order of 1 $g.m^{-3}.min^{-1}$), so that this does not reach concentrations that are toxic for microalgae. The supply of carbon and the controlling of the pH level can be ensured by the injection of CO_2. This can be used for the measurement of the pH, or fixed so as to represent a constant percentage of the gaseous mix of air–CO_2. Finally, these tubular systems, built on a larger scale, exist in some private companies over hundred of square meters of useful cultures.

2.4. Research in support of the development of the branch

2.4.1. *Omics*

Like with research in the domains of plant biology for agronomy, or microbiology for biotechnology, the acquisition of knowledge on microalgae species, whether they are models or exploited, allows the improvement of their use and the discovery of new fields of biotechnological applications. The rise of molecular biology in the second half of the 20th Century, and the technological sequencing revolution that started in the 2000s, have transformed the world of biological research. Work on microalgae using genomic and post-genomic approaches majorly involved the species *C. reinhardtii*, a model chlorophyte species, of interest for its taxonomic proximity with upright plants [HAR 01]. The genome of this species has been sequenced [MER 07]. The numerous tools and molecular approaches (RNA chips, antibodies, RNA interference, RNA seq, proteomics, etc.) developed using this model unicellular species have allowed for a better understanding of the metabolisms and cellular mechanisms of photosynthetic eukaryotic organisms. In 2013, 27 microalgae models were entirely sequenced, spanning a considerable amount of biodiversity, which included chlorophyceae, diatoms, a rhodophyceae, a cryptophyte and a haptophyte[4]. The structures and the genic composition of the genomes of microalgae proved to be complex. Their sizes vary strongly in function of the species: 12.6 Mpb for the chlorophyceae *Ostreococcus tauri*; 168 Mpb for the haptophyte *Emiliania huxleyi*; 10,000 Mpb estimated for the dynophyceae *Karenia brevis*. The genomic data obtained on these species are completed by the transcriptomes of various species of phylogenomic and biotechnological interest, both published and in the middle of being sequenced. All of this genomic data has led, on the one hand, to the better description of phenomena of endosymbiosis and evolution which give this extraordinary to microalgae, and on the other hand, to establish a base of biological knowledge for ecological and functional studies.

The genomic analyses within biotechnology mainly involve the exploitation of lipids and the secondary metabolites of microalgae. The exploration of the genomes and transcriptomes of species that produce polyunsaturated long chain fatty acids such as *Pseudochattonella farcimen*, *Myrmecia incisa* and *Nannochloropsis oceanica*, has led to the identification of the enzymes involved in the biosynthesis of these lipids of nutritional interest [YU 11]. In the same manner, many teams have set up genomic analyses to identify *in silico* genes of interest for the biosynthesis of secondary metabolites of high biotechnological potential such as phycotoxines, pigments or phycocolloids [ZHA 11]. The exploration of genomes is a promising strategy, not only for the identification of pathways for the biosynthesis of natural products of interest, but also for the identification of new bioactive compounds [WIN 11].

4 See http://wwz.ifremer.fr/pba.

Figure 2.9. *Culture of the microalgae Nannochloropsis salina in a round-bottom flask (photo Kaas/Ifremer)*

Recently, in response to the rising popularity of third generation biofuels, several research teams have developed genomic, transcriptomic and proteomic approaches to characterize metabolic pathways, and, more precisely, the biosynthetic pathways of triacylglycerides in oleaginous microalgae such as *Dunaliella tertiolecta*, *B. braunii*, *H. pluvialis*, *Chlorella vulgaris* or *Eustigmatos cf. polyphem* [WAN 12]. Generally speaking, the main metabolic pathways involved in the metabolism of lipids present homologies with upright plants, but certain pathways specific for microalgae have also been identified [FAN 11]. Furthermore, the regulatory mechanisms of lipid synthesis are currently mostly unknown in microalgae.

To this day, the scientific community generates masses of genomic data, which constitutes a base for the optimization of the exploitation of microalgae and for the discovery of new biotechnological potential. Less than 50% of the genes of microalgae studied can be identified as homologous with known genes in other organisms. The functions of most of them have never been confirmed. While efforts have been made in model algae, and notably in the chlorophyceae *C. reinhardtii*, the taxonomic distances between the different phyla limit the functional exploration of the genomes. The study of the function of genes and proteins and of the regulatory mechanisms at the levels of transcription, translation and post-translation is a promising perspective for the exploration of genomic data for biotechnology.

2.4.2. *Species improvement*

Microalgae represent a still largely unexplored world, and the few commercial cultures that exist rely on limited number of species. All of the species cultivated are "as discovered", as the microalgae have never been selected, but only deposited in banks and transferred from tube to tube. As in the case of the strategies developed in agriculture, and depending on the biology of the species, the access to diversity and the mode of reproduction, the improvement methods that can be put into place are varied. Generally, four main selection strategies can be identified: (1) by genetic engineering, (2) by controlled crossing, (3) individual screening, or (4) by artificial and controlled induction of genetic variability [MUR 07].

The genetic engineering approach consists of introducing a gene or an allele of interest coming from one organism into another, whether it is of the same species or not. This technique has the advantage of only introducing the gene or the allele of interest, in an already cultivated strain. In most cases, the transfer concerns one gene, with three or four at most. In microalgae, this technique is largely used in fundamental research, and their applications have been reviewed [CAD 12] and are subject to several patents [BAR 10, CAD 09]. A large part of the families of microalgae now have transformed representatives. The diversity mentioned above makes adapting techniques necessary, in function, for example, of the fragilities of membranes or of the specificity of gene-regulating sequences (promoters, GC content, etc.). In the context of the elaboration of complex pharmaceutical molecules, this strategy is already a reality and involves a number of companies around the world [BAR 10, CAD 09, CAR 10, JON 12,

LEJ 10, TRA 13]. In the context of "biofuels", some biosynthesis pathways are targeted and fully identified. Modifications in these labyrinths remain hypothetical and risky, but this choice has still been made by some of those in the industry who see this path as the most promising.

The strategy of selective breeding by crossing individuals consists of choosing two parents for their aptitudes and exploiting their progeny as best possible. Different methods can be used to characterize and exploit this progeny according to the objective and the reproductive mode of the species (genealogical, backcrossing, etc.) [TES 10]. With regards to microalgae, the cycle and the modes of reproduction are still poorly understood, with the exception of some model species like *C. reinhardtii* [CHI 67], *Thalassiosira pseudonana* and *Phaeodactylum tricornutum* [CHE 11]. This lack of knowledge makes it difficult to use crosses as selection procedures in many microalgae and this selection strategy will be, in the future, very dependent on the elucidation of the varied reproductive cycles, specific for each group.

The birth of genetics in the 20th Century, and more generally the birth of "omics" in the present day (genomics, transcriptomics, proteomics, etc.), enables the establishment of new tools such as marker-assisted selection [LAN 11]. Once a population demonstrating high diversity is found, a screening-based approach is usually carried out and consists of selecting one or several individuals with characteristics of interest observed in a population. In microalgae, these approaches concern some characteristics of the phenotype [BOU 12]. The difficulty of the establishment of high-speed phenotyping techniques currently limits the number of characteristics that can be screened for. New molecular marker screening methods (marker-assisted selection) have been developed. They could allow the selecting of strains based on their genes/alleles of interest to avoid high-speeding phenotyping [COL 08]. This type of approach requires a relatively large amount of knowledge on the studied species [XU 08]. These last few years, the capabilities and the cost of genotyping have permitted the analysis of several hundred individuals at the same time for a large number of markers [VAR 09]. Several strategies have appeared to identify the markers linked to phenotypes, protein research, transcripts, genome regions (QTL), mutations (QTN), genes or alleles of interest [FOU 09] and could be, in part, transferable to microalgae. Currently, 27 microalgae genomes have been sequenced as well as several dozen of transcriptomes [CAD 12], thus providing a considerable amount of data.

As for plants, the exploitation of biodiversity requires the conduction of prospective campaigns, conservation strategies and the establishment of collections [ULU 11]. Currently, collections of microalgae exist but they usually only cover a small number of the strains present in each type of algae and look more at interspecies diversity. For examples see the *Roscoff Culture Collection* (www.sb-roscoff.fr/Phyto/RCC/index.php), the *Australian National Algae Culture Collection* (www.marine.csiro.au/algaedb/default.htm), the *Culture Collection of Algae and Protozoa* (www.ccap.ac.uk/index.htm) or the *National Center for Marine Algae and Microbiota* (https://ncma.bigelow.org/). However, initial studies on a small number of microalgae species show that there is a large amount of intra-species diversity, whether at the genetic level [EVA 09] and/or at the level of the phenotype [MEN 12]. The exploitation of biodiversity in microalgae therefore seems to be an attractive potential source of improvement.

Unlike the selective breeding of plants and animals, which started in Antiquity, the first programs for the selection of microalgae have only started recently, but benefit from methods and tools developed in other organisms. This transfer of knowledge could allow for the retrieval of improved microalgae strains relatively quickly.

Methods based on isolation have already allowed for the effective selection of microalgae strains [MOL 95], knowing that the use of chemical mutagen agents or even UV rays can induce changes leading to improved performance [BON 11], even if this generated diversity remains limited compared to diversity obtained from crossing [GRA 05]. [SHA 91] used UV irradiation to mutate and then select strains of *Dunaliella bardawil* that are rich in ß-carotene. By using a similar process, [ALO 96] managed to increase the EPA content of the diatom *P. tricornutum* by 37%. More recently, [MEI 03] succeeded in increasing the production of EPA and DHA by 33% in *Pavlova lutheri* by using UV rays. Chemical mutagenesis agents, such as *ethyl methane sulfonate* (EMS), have also been used successfully to improve the production of a variety of compounds. In this way, [CHA 06] increased the production of EPA in *Nannochloropsis oculata* and [MEN 08] selected a strain that hyperproduced carotenoids in *D. salina*. The selection of cells can be operated manually by limit dilution or by micromanipulation [VIG 12]. However, automatic sorting techniques are a lot quicker, resulting in an increase in the number of cell sorted and, as a result, in a quicker selection of the desired population. The combination of a fluorescent vital marker and sorting by flow cytometry has been described for the selection of high lipid contents in *Tetraselmis suecica* [MON 10], *Nannochloropsis sp.* [DOA 12]. The latters added a chemical mutation step (CMS), while the combination of "UV mutagenesis" with an automation of the sorting in the flow cytometer to select for lipid-producing cells has been seen in the isochrysis *T. lutea* [BEN 13], commonly cultivated in aquaculture. This strategy has led to a doubling of the total fatty acid content in this microalgae [BOU 12], and this in a stable manner over several years. These works have led to a patent in France of the first improved microalgae strain, in 2011 [ROU 11].

2.5. Conclusion

Thousands of microalgae are now on shelves around the world and thousands remain to be discovered. Some of the main algal libraries provide for the many research teams that specialize in all the domains of application cited above. These libraries are distributed over the globe (see *supra*). Despite this, the quest for new species is the objective of many projects, of various sizes. Aside from the laboratories that isolate one species for particular characteristics, large and ambitious programs finance prospective campaigns all over the world. One of the first was led by Craig Venter: on the ship Sorcerer II during the expedition: *Global Ocean Sampling Expedition* (GOS) (www.sorcerer2expedition.org/). The same idea is behind all of the Tara cruises (http://oceans.taraexpeditions.org/), recently completed by the Oceanomics program, which aims to understand the biocomplexity and the biotechnological potential of oceanic plankton (http://oceans.taraexpeditions.org/fr/). This leads to the notion of metagenomics, reflecting the global sequencing of masses of water, "blindly", for which the assembly and the interpretation would be carried out *a posteriori in silico*.

The estimation of the number of species will become more accurate in this way. In the mean time, we can read the review by Michael Guiry [GUI 12] as the most informative and balanced, which bases itself significantly on the site Algaebase: www.algaebase.org/.

Out of the thousands of species known currently, only a few dozen are exploited. Surprisingly, all of the species are "as discovered", i.e. in the same state that they were in when discovered, sometimes dozens of years ago, and placed in a tube. In 2013, less than five publications were on structured works of selection attempting to imitate the world of terrestrial plants so as to improve performances.

In terms of the valorization of biotechnology, microalgae present a series of advantages that are reflected in the high levels of global investment. The yields of dry biomass per square meter and per day are higher than for many terrestrial plants. It is true that the surfaces that are needed for cultivation in no way need to be cultivatable lands; on the contrary it must be noted that industrial cultures of marine microalgae would preserve freshwater reserves.

As for all plants, the vital sources come from fertilizer and carbon. These fertilizers can be supplied by urban or agricultural effluents or administered chemically. This is also true for carbon. We can imagine the use of gas taken from the bottom of industrial chimneys or the provision of pure gas from bottles. The culture of microalgae would then fit into an industrial environment as a system for the fixation of nitrogen, phosphorus, CO_2 or even heavy metals, for later use or valorization.

We are currently on the verge of appropriating an original, extraordinary diverse plant that remains virtually untouched. This plant source will complement terrestrial stocks, and its exploitation will profit from modern agricultural knowledge.

2.6. Bibliography

[ABU 09] ABUGHAZALEH A.A., POTU R.B., IBRAHIM S., "Short communication: The effect of substituting fish oil in dairy cow diets with docosahexaenoic acid-micro algae on milk composition and fatty acids profile", *Journal of Dairy Science*, vol. 92, pp. 6156–6159, 2009.

[ALO 96] ALONSO D.L., SEGURA DEL CASTILLO C.I., MOLINA GRIMA E., *et al.*, "First insight into improvement of eicosapentaenoic acid content in *Phaeodactylum tricornutum* (Bacillariophyceae) by induced mutagenesis", *Journal of Phycology*, vol. 32, pp. 339–345, 1996.

[APT 99] APT K.E., BEHRENS P.W., "Commercial developments in microalgal biotechnology", *Journal of Phycology*, vol. 35, pp. 215–226, 1999.

[BAI 86] BAILEY J.E., OLLIS D.F., *Biochemical Engineering Fundamentals*, McGraw-Hill, New York, 1986.

[BAO 07] BAO Z., WEATHERSPOON M.R., SHIAN S., *et al.*, "Chemical reduction of three-dimensional silica micro-assemblies into microporous silicon replicas", *Nature*, vol. 446, pp. 172–175, 2007.

[BAR 10] BARDOR M., LOUVET R., SAINT-JEAN B., *et al.*, N-Glycosylation in transformed *Phaeodactylum tricornutum*, brevet: WO2012013337A1, 2010.

[BEC 07] BECKER W., "Microalgae in Human and Animal Nutrition", in RICHMOND A. (ed.), *Handbook of Microalgal Culture: Biotechnology and Applied Phycology*, Blackwell Publishing, Oxford, 2007.

[BEN 13] BENDIF E.M., PROBERT I., SCHROEDER D.C., *et al.*, "On the description of *Tisochrysis lutea gen. nov. sp. nov.* and *Isochrysis nuda* sp. *nov.* in the Isochrysidales, and the transfer of Dicrateria to the Prymnesiales (Haptophyta)", *Journal of Applied Phycology*, vol. 25, pp. 1763–1776, 2013.

[BEN 97] BENEMANN J.R., "CO_2 mitigation with microalgae systems", *Energy Conversion and Management*, vol. 38, pp. 475–479, 1997.

[BHA 02] BHATTI S., HUERTAS I.E., COLMAN B., "Acquisition of inorganic carbon by the marine haptophyte *Isochrysis galbana* (prymnesiophyceae)1", *Journal of Phycology*, vol. 38, pp. 914–921, 2002.

[BLA 05] BLACKMAN F.F., "Optima and Limiting Factors", *Annals of Botany*, vol. 19, pp. 281–296, 1905.

[BON 11] BONENTE G., FORMIGHIERI C., MANTELLI M., *et al.*, "Mutagenesis and phenotypic selection as a strategy toward domestication of *Chlamydomonas reinhardtii* strains for improved performance in photobioreactors", *Photosynthesis Research*, vol. 108, pp. 107–120, 2011.

[BOU 10] BOUGARAN G., BERNARD O., SCIANDRA A., "Modeling continuous cultures of microalgae colimited by nitrogen and phosphorus", *Journal of Theoretical Biology*, vol. 265, pp. 443–454, 2010.

[BOU 03] BOUGARAN G., LE DÉAN L., LUKOMSKA E., *et al.*, "Transient initial phase in continuous culture of *Isochrysis galbana affinis* Tahiti", *Aquatic Living Resources*, vol. 16, pp. 389–394, 2003.

[BOU 12] BOUGARAN G., ROUXEL C., DUBOIS N., *et al.*, "Enhancement of neutral lipid productivity in the microalga *Isochrysis affinis galbana* (T-Iso) by a mutation-selection procedure", *Biotechnology Bioengineering*, vol. 109, pp. 2737–2745, 2012.

[BRI 04] BRITTON G., LIAAEN-JENSEN S., PFANDER H.P., *Handbook of Carotenoids*, Springer, Vienna, 2004.

[BUI 03] BUITENHUIS E.T., TIMMERMANS K.R., DE BAAR H.J., "Zinc-bicarbonate colimitation of *Emiliania huxley*", *Limnology and Oceanography*, vol. 48, pp. 1575–1582, 2003.

[BUM 11] BUMBAK F., COOK S., ZACHLEDER V., *et al.*, "Best practices in heterotrophic high-cell-density microalgal processes: achievements, potential and possible limitations", *Applied Microbiology and Biotechnology*, vol. 91, pp. 31–46, 2011.

[BUO 12] BUONO S., LANGELLOTTI A.L., MARTELLO A., *et al.*, "Biological activities of dermatological interest by the water extract of the microalga *Botryococcus braunii*", *Archives of Dermatological Research*, vol. 304, pp. 755–764, 2012.

[CAD 08a] CADORET J.P., BARDOR M., LEROUGE P., *et al.*, "Les microalgues : Usines cellulaires productrices de molécules commerciales recombinantes", *Médecine/Sciences*, vol. 24, pp. 375–382, 2008.

[CAD 08b] CADORET J.P., BERNARD O., "La production de biocarburant lipidique avec des microalgues: promesses et défis", *Journal Société Biologie*, vol. 202, pp. 201–211, 2008.

[CAD 09] CADORET J.P., CARLIER A., BUREL C., *et al.*, Production of glysosylated proteins in microalgae, PCT/EP/2009/051672, 2009.

[CAD 12] CADORET J.P., GARNIER M., SAINT-JEAN B., "Microalgae, functional genomics and biotechnology", in PIGANEAU G. (ed.), *Advances in Botanical Research*, Academic Press, New York, 2012.

[CAD 48] CADY J.B., MORGAN W.S., "Treatment of chronic ulcers with chlorophyll: review of a series of fifty cases", *American Journal of Surgery*, vol. 75, pp. 562–569, 1948.

[CAR 10] CARLIER A., MICHEL R., DUFOURMANTEL N., *et al.*, Production of high mannose glycosylated proteins stored in the plastid of microalgae, EP10016162.9, 2010.

[CHA 06] CHATURVEDI R., FUJITA Y., "Isolation of enhanced eicosapentaenoic acid producing mutants of *Nannochloropsis oculata* using ethyl methane sulfonate induced mutagenesis techniques and their characterization at mRNA transcript level", *Phycological Research*, vol. 54, pp. 208–219, 2006.

[CHE 11] CHEPURNOV V.A., CHAERLE P., ROEF L., *et al.*, "Classical breeding in diatoms: scientific background and practical perspectives", in SECKBACH J., KOCIOLEK P. (eds.), *Cellular Origin, Life in Extreme Habitats and Astrobiology*, Springer, Amsterdam, 2011.

[CHI 67] CHIANG K.S., SUEOKA N., "Replication of chromosomal and cytoplasmic DNA during mitosis and meiosis in the eucaryote *Chlamydomonas reinhardi*", *Journal of Cellular Physiology*, vol. 70, pp. 89–112, 1967.

[CHR 13] CHRISTAKI E., BONOS E., GIANNENAS I., *et al.*, "Functional properties of carotenoids originating from algae", *Journal of the Science of Food and Agriculture*, vol. 93, pp. 5–11, 2013.

[COL 08] COLLARD B.C., MACKILL D.J., "Marker-assisted selection: an approach for precision plant breeding in the twenty-first century", *Philosophical Transaction of the Royal Society B: Biological Science*, vol. 363, pp. 557–572, 2008.

[COM 94] COMBRES C., LALIBERTÉ G., REYSSAC J.S., DE LA NOÜE J., "Effect of acetate on growth and ammonium uptake in the microalgae *Scenedesmus obliquus*", *Plant Physiology*, vol. 91, pp. 729–734, 1994.

[CON 09] CONVERTI A., CASAZZA A.A., ORTIZ E.Y., *et al.*, "Effect of temperature and nitrogen concentration on the growth and lipid content of *Nannochloropsis oculata* and *Chlorella vulgaris* for biodiesel production", *Chemical Engineering and Processing: Process Intensification*, vol. 48, pp. 1146–1151, 2009.

[COO 11] COONEY M.J., YOUNG G., PATE R., "Bio-oil from photosynthetic microalgae: case study", *Bioresource Technology*, vol. 102, pp. 166–177, 2011.

[CSO 99] CSOGOR Z., MELGAR D., SCHMIDT K., *et al.*, "Production and particle characterization of the frustules of *Cyclotella cryptica* in comparison with siliceous earth", *Journal of Biotechnology: Biotechnological Aspects of Marine Sponges*, vol. 70, pp. 71–75, 1999.

[DEL 07] DEL CAMPO J.A., GARCÍA-GONZÁLEZ M., GUERRERO M.G., "Outdoor cultivation of microalgae for carotenoid production: current state and perspectives", *Applied Microbiology and Biotechnology*, vol. 74, pp. 1163–1174, 2007.

[DIC 87] DICKSON A.G., MILLERO F.J., "A comparison of the equilibrium constants for the dissociation of carbonic acid in seawater media", *Deep Sea Research Part I: Oceanographic Research Papers*, vol. 34, pp. 1733–1743, 1987.

[DOA 12] DOAN T.T.Y., OBBARD J.P., "Enhanced intracellular lipid in *Nannochloropsis* sp. via random mutagenesis and flow cytometric cell sorting", *Algal Research*, vol. 1, pp. 17–21, 2012.

[DRO 68] DROOP M.R., "Vitamin B12 and marine ecology. IV. The kinetics of uptake, growth and inhibition in *Monochrysis Lutheri*", *Journal of the Marine Biological Association of the United Kingdom*, vol. 48, pp. 689–733, 1968.

[EIX 05] EIXLER S., SELIG U., KARSTEN U., "Extraction and detection methods for polyphosphate storage in autotrophic planktonic organisms", *Hydrobiologia*, vol. 533, pp. 135–143, 2005.

[ELR 85] ELRIFI I.R., TURPIN D.H., "Steady-state luxury consumption and the concept of optimum nutrient ratios: a study with phosphate and nitrate limited *Selenastrum minutum* (chlorophyta)", *Journal of Phycology*, vol. 21, pp. 592–602, 1985.

[EPP 72] EPPLEY R.W., "Temperature and phytoplankton growth in the sea", *Fish Bulletin,* vol. 70, 1972.

[EVA 09] EVANS K.M., CHEPURNOV V.A., SLUIMAN H.J., *et al.*, "Highly differentiated populations of the freshwater diatom *Sellaphora capitata* suggest limited dispersal and opportunities for allopatric speciation", *Protist,* vol. 160, pp. 386–396, 2009.

[FAL 75] FALKOWSKI P.G., "Nitrate uptake in marine phytoplankton: comparison of half-saturation constants from seven species", *Limnology and Oceanography*, vol. 20, pp. 412–417, 1975.

[FAN 11] FAN J., ANDRE C., XU C., "A chloroplast pathway for the de novo biosynthesis of triacylglycerol in *Chlamydomonas reinhardtii*", *FEBS Letters*, vol. 585, no. 12, pp. 1985–1991, 2011.

[FAR 06] FARINEAU J., MOROT-GAUDRY J.F., *La Photosynthèse: processus physiques, moléculaires et physiologiques*, Quae, Versailles, 2006.

[FIN 10] FINAZZI G., MOREAU H., BOWLER C., "Genomic insights into photosynthesis in eukaryotic phytoplankton", *Trends in Plant Science*, vol. 15, pp. 565–572, 2010.

[FOU 09] FOURNIER-LEVEL A., LE CUNFF L., GOMEZ C., *et al.*, "Quantitative genetic bases of anthocyanin variation in grape (*Vitis vinifera L. ssp. sativa*) berry: a quantitative trait locus to quantitative trait nucleotide integrated study", *Genetics*, vol. 183, pp. 1127–1139, 2009.

[GAD 09] GADD G.M., "Biosorption: critical review of scientific rationale, environmental importance and significance for pollution treatment", *Journal of Chemical Technology and Biotechnology*, vol. 84, pp. 13–28, 2009.

[GAG 12] GAGEZ A.L., THIERY V., PASQUET V., *et al.*, "Epoxycarotenoids and cancer", *Review Current Bioactive Compounds*, vol. 8, pp. 109–141, 2012.

[GEI 02] GEIDER R., LA ROCHE J., "Redfield revisited: variability of C:N:P in marine microalgae and its biochemical basis", *European Journal of Phycology*, vol. 37, pp. 1–17, 2002.

[GEI 87] GEIDER R.J., "Light and temperature dependence of the carbon to chlorophyll a ratio in microalgae and cyanobacteria: implications for physiology and growth of phytoplankton", *New Phytologist*, vol. 106, pp. 1–34, 1987.

[GLA 94] GLADUE R.M., MAXEY J.E., "Microalgal feeds for aquaculture", *Journal of Applied Phycology*, vol. 6, pp. 131–141, 1994.

[GOR 09] GORDON R., LOSIC D., TIFFANY M.A., *et al.*, "The glass menagerie: diatoms for novel applications in nanotechnology", *Trends in Biotechnology*, vol. 27, pp. 116–127, 2009.

[GOU 08] GOUVEIA L., COUTINHO C., MENDONCA E., *et al.*, "Functional biscuits with PUFA-omega 3 from *Isochrysis galbana*", *Journal of the Science of Food and Agriculture*, vol. 88, pp. 891–896, 2008.

[GRA 05] GRAPPUTO A., KUMPULAINEN T., MAPPES J., *et al.*, "Genetic diversity in populations of asexual and sexual bag worm moths (Lepidoptera: Psychidae)", *BMC Ecology*, vol. 5, p. 5, 2005.

[GRO 03] GROBBELAAR J.U., KURANO N., "Use of photoacclimation in the design of a novel photobioreactor to achieve high yields in algal mass cultivation", *Journal of Applied Phycology*, vol. 15, pp. 121–126, 2003.

[GUE 03] GUERIN M., HUNTLEY M.E., OLAIZOLA M., "*Haematococcus* astaxanthin: applications for human health and nutrition", *Trends in Biotechnology*, vol. 21, pp. 210–216, 2003.

[GUI 12] GUIRY M.D., "How many species of algae are there?", *Journal of Phycology*, vol. 48, pp. 1057–1063, 2012.

[HAI 13] HAIMOVICH-DAYAN M., GARFINKEL N., EWE D., *et al.*, "The role of C_4 metabolism in the marine diatom *Phaeodactylum tricornutum*", *New Phytologist*, vol. 197, pp. 177–185, 2013.

[HAR 01] HARRIS E.H., "*Chlamydomonas* as a model organism", *Annual Review of Plant Physiology and Plant Molecular Biology*, vol. 52, pp. 363–406, 2001.

[HEM 11] HEMPEL F., LAU J., KLINGL A., *et al.*, "Algae as protein factories: expression of a human antibody and the respective antigen in the diatom *Phaeodactylum tricornutum*", *PLoS One*, vol. 6, p. e28424, 2011.

[HER 11] HEREDIA-ARROYO T., WEI W., RUAN R., *et al.*, "Mixotrophic cultivation of *Chlorella vulgaris* and its potential application for the oil accumulation from non-sugar materials", *Biomass Bioenergy*, vol. 35, pp. 2245–2253, 2011.

[HOR 51] HORWITH B., "Role of chlorophyll in proctology", *American Journal of Surgery*, vol. 81, pp. 81–84, 1951.

[JEN 01] JENSEN G.S., GINSBERG D., DRAPEAU C., "Blue-green algae as an immuno-enhancer and biomodulator", *Journal of the American Nutraceutical Association*, vol. 3, pp. 24–30, 2001.

[JON 12] JONES C.S., LUONG T., HANNON M., *et al.*, "Heterologous expression of the C-terminal antigenic domain of the malaria vaccine candidate Pfs48/45 in the green algae *Chlamydomonas reinhardtii*", *Applied Microbiology and Biotechnology*, vol. 97, pp. 1987–1995, 2012.

[KEE 99] KEELING P.J., DEANE J.A., HINK-SCHAUER C., *et al.*, "The secondary endosymbiont of the cryptomonad *Guillardia theta* contains Alpha-, Beta-, and Gamma-tubulin Genes", *Molecular Biology and Evolution*, vol. 16, pp. 1308–1313, 1999.

[KIR 94] KIRK J.T.O., *Light and Photosynthesis in Aquatic Ecosystems*, 2nd ed., Cambridge University Press, Cambridge, 1994.

[KO 12] KO S.C., KIM D., JEON Y.J., "Protective effect of a novel antioxidative peptide purified from a marine *Chlorella ellipsoidea* protein against free radical-induced oxidative stress", *Food and Chemical Toxicology*, vol. 50, pp. 2294–2302, 2012.

[KUE 65] KUENZLER E.J., PERRAS J.P., "Phosphatases of marine algae", *Biological Bulletin*, vol. 128, pp. 271–284, 1965.

[LAL 93] LALIBERTÉ G., DE LA NOÜE J., "Auto-, hetero-, and mixotrophic growth of *chlamydomonas humicola* (cmloroimiyckak) on acetate1", *Journal of Phycology*, vol. 29, pp. 612–620, 1993.

[LAN 11] LANGRIDGE P., FLEURY D., "Making the most of "omics" for crop breeding", *Trends in Biotechnology*, vol. 29, pp. 33–40, 2011.

[LEE 99] LEE C.G., "Calculation of light penetration depth in photobioreactors", *Biotechnology and Bioprocess Engineering*, vol. 4, pp. 78–81, 1999.

[LEE 97] LEE Y.K., "Commercial production of microalgae in the Asia-Pacific rim", *Journal of Applied Phycology*, vol. 9, pp. 403–411, 1997.

[LEJ 10] LEJEUNE A., MICHEL R., CADORET J.P., *et al.*, Secretion of recombinant polypeptides in the extracellular medium of diatoms, EP10013808.0, 2010.

[LIA 09] LIANG Y., SARKANY N., CUI Y., "Biomass and lipid productivities of *Chlorella vulgaris* under autotrophic, heterotrophic and mixotrophic growth conditions", *Biotechnology Letters*, vol. 31, pp. 1043–1049, 2009.

[LOM 00] LOMAS M.W., GLIBERT P.M., "Comparisons of nitrate uptake, storage, and reduction in marine diatoms and flagellates", *Journal of Phycology*, vol. 36, pp. 903–913, 2000.

[LUA 11] LUANGPIPAT T., BEATTIE I.R., CHISTI Y., *et al.*, "Gold nanoparticles produced in a microalga", *Journal of Nanoparticle Research*, vol. 13, pp. 6439–6445, 2011.

[MAC 02] MACINTYRE H.L., KANA T.M., ANNING T., *et al.*, "Photoacclimation of photosynthesis irradiance response curves and photosynthetic pigments in microalgae and cyanobacteria1", *Journal of Phycology*, vol. 38, pp. 17–38, 2002.

[MAR 13] MARCHETTI J., BOUGARAN G., JAUFFRAIS T., *et al.*, "Effects of blue light on the biochemical composition and photosynthetic activity of *Isochrysis* sp. (T-iso)", *Journal of Applied Phycology*, vol. 25, pp. 109–119, 2013.

[MAR 12] MARCHETTI J., BOUGARAN G., LE DEAN L., *et al.*, "Optimizing conditions for the continuous culture of *Isochrysis* affinis *galbana* relevant to commercial hatcheries", *Aquaculture*, vol. 326–329, pp. 106–115, 2012.

[MEI 03] MEIRELES L., GUEDES A.C., MALCATA F.X., "Increase of the yields of eicosapentaenoic and docosahexaenoic acids by the microalga *Pavlova lutheri* following random mutagenesis", *Biotechnology Bioengineering*, vol. 81, pp. 50–55, 2003.

[MEL 09] MELIS A., "Solar energy conversion efficiencies in photosynthesis: minimizing the chlorophyll antennae to maximize efficiency", *Plant Science*, vol. 177, pp. 272–280, 2009.

[MEN 09] MENDES A., REIS A., VASCONCELOS R., *et al.*, "*Crypthecodinium cohnii* with emphasis on DHA production: a review", *Journal of Applied Phycology*, vol. 21, pp. 199–214, 2009.

[MEN 12] MENDOZA H., DE LA JARA A., FREIJANES K., *et al.*, "Quick estimation of intraspecific variation of fatty acid composition in *Dunaliella salina* using flow cytometry and Nile Red", *Journal of Applied Phycology*, vol. 24, pp. 1237–1243, 2012.

[MEN 08] MENDOZA H., DE LA JARA A., FREIJANES K., *et al.*, "Characterization od *Dunaliella salina* strains by flow cytometry: a new approach to select carotenoid hyperproducing strains", *Electronic Journal Biotechnology*, vol. 11, pp. 1–13, 2008.

[MER 07] MERCHANT S.S., "The *Chlamydomonas* genome reveals the evolution of key animal and plant functions", *Science*, vol. 318, pp. 245–250, 2007.

[MOL 95] MOLINA GRIMA E., SANCHEZ PEREZ J.A., GARCIA CAMACHO F., *et al.*, "The production of polyunsaturated fatty acids by microalgae: from strain selection to product purification", *Process Biochemistry*, vol. 30, pp. 711–719, 1995.

[MON 10] MONTERO M., ARISTIZABAL M., GARCIA REINA G., "Isolation of high-lipid content strains of the marine microalga *Tetraselmis suecica* for biodiesel production by flow cytometry and single-cell sorting", *Journal of Applied Phycology*, pp. 1–5, 2010.

[MOR 12] MOREIRA I. DE O., PASSOS T.S., CHIAPINNI C., *et al.*, "Colour evaluation of a phycobiliprotein-rich extract obtained from *Nostoc* PCC9205 in acidic solutions and yogurt", *Journal of the Science of Food and Agriculture*, vol. 92, pp. 598–605, 2012.

[MUR 07] MURPHY D., *Plant Breeding and Biotechnology*, Cambridge University Press, Cambridge, 2007.

[OGB 03] OGBONNA J.C., "Photobioreactors", *Recent Advance in Marine Biotechnology*, vol. 9, pp. 315–348, 2003.

[OLI 07] OLIVO E., Conception et étude d'un photobioréacteur pour la production en continu de microalgues en écloseries aquacoles, University of Nantes, 2007.

[PAH 09] PAHLOW M., OSCHLIES A., "Chain model of phytoplankton P, N and light colimitation", *Marine Ecology Progress Series*, vol. 376, pp. 69–83, 2009.

[PAU 95] PAULY D., CHRISTENSEN V., "Primary production required to sustain global fisheries", *Nature*, vol. 374, pp. 255–257, 1995.

[PER 11] PEREZ-GARCIA O., ESCALANTE F.M.E., DE-BASHAN L.E., *et al.*, "Heterotrophic cultures of microalgae: metabolism and potential products", *Water Research*, vol. 45, pp. 11–36, 2011.

[POT 10] POTVIN G., ZHANG Z., "Strategies for high-level recombinant protein expression in transgenic microalgae: a review", *Biotechnology Advances*, vol. 28, pp. 910–918, 2010.

[PUL 01] PULZ O., "Photobioreactors: production systems for phototrophic microorganisms", *Applied Microbiology and Biotechnology*, vol. 57, no. 3, pp. 287–293, 2001.

[PUL 04] PULZ O., GROSS W., "Valuable products from biotechnology of microalgae", *Applied Microbiology and Biotechnology*, vol. 65, pp. 635–648, 2004.

[RAP 13] RAPOSO M., DE MORAIS R., BERNARDO DE MORAIS A., "Bioactivity and applications of sulphated polysaccharides from marine microalgae", *Marine Drugs*, vol. 11, pp. 233–252, 2013.

[RAV 99] RAVEN J.A., FALKOWSKI P.G., "Oceanic sinks for atmospheric CO sub(2)", *Plant Cell and Environment*, vol. 22, pp. 741–755, 1999.

[RED 34] REDFIELD A.C., On the proportions of organic derivatives in sea water and their relation to the composition of plankton, University Press of Liverpool, Liverpool, 1934.

[REX 02] REXACH J., LLAMAS A., FERNANDEZ E., *et al.*, "The activity of the high-affinity nitrate transport system I (NRT2;1, NAR2) is responsible for the efficient signalling of nitrate assimilation genes in *Chlamydomonas reinhardtii*", *Planta*, vol. 215, pp. 606–611, 2002.

[RHE 78] RHEE G.Y., "Effects of N:P atomic ratios and nitrate limitation on algal growth, cell composition, and nitrate uptake", *Limnology and Oceanography*, vol. 23, pp. 10–24, 1978.

[RIC 08] RICHMOND A., *Handbook of Microalgal Culture: Biotechnology and Applied Phycology*, John Wiley & Sons, New York, 2008.

[RIC 06] RICO-VILLA B., LE COZ J.R., MINGANT C., *et al.*, "Influence of phytoplankton diet mixtures on microalgae consumption, larval development and settlement of the Pacific oyster *Crassostrea gigas* (Thunberg)", *Aquaculture*, vol. 256, pp. 377–388, 2006.

[ROB 09] ROBERTSON C.A., EVANS D.H., ABRAHAMSE H., "Photodynamic therapy (PDT): a short review on cellular mechanisms and cancer research applications for PDT", *Journal of Photochemistry and Photobiology*, vol. 96, pp. 1–8, 2009.

[ROS 12] RÖSCH C., POSTEN C., "Challenges and perspectives of microalgae production", *Technikfolgenabschätzung*, pp. 5–16, 2012.

[ROS 10] ROSELLO SASTRE R., POSTEN C., "Die vielfältige Anwendung von Mikroalgen als nachwachsende Rohstoffe", *Chemie Ingenieur Technik*, vol. 82, pp. 1925–1939, 2010.

[ROU 11] ROUXEL C., BOUGARAN G., DOULIN-GROUAS S., *et al.*, Novel isochrysis sp tahitian clone and uses therefore, EP 11006712.1, 2011.

[SAI 08] SAITO M.A., GOEPFERT T.J., RITT J.T., "Some thoughts on the concept of colimitation: three definitions and the importance of bioavailability", *Limnology and Oceanography*, vol. 53, pp. 276–290, 2008.

[SCA 10] SCARSELLA M., BELOTTI G., DE FILIPPIS P., *et al.*, "Study on the optimal growing conditions of *Chlorella vulgaris* in bubble column photobioreactors", *Chemical Engineering Transactions*, vol. 20, pp. 85–90, 2010.

[SCH 04] SCHMIDT F.R., "Recombinant expression systems in the pharmaceutical industry", *Applied Microbiology and Biotechnology*, vol. 65, no. 4, pp. 363–372, 2004.

[SCH 13] SCHWENZFEIER A., HELBIG A., WIERENGA P.A., *et al.*, "Emulsion properties of algae soluble protein isolate from *Tetraselmis* sp.", *Food Hydrocolloids*, vol. 30, pp. 258–263, 2013.

[SCR 99] SCRIBAN R., *Biotechnologie*, Lavoisier Tec & Doc, Paris, 1999.

[SHA 91] SHAISH A., BEN-AMOTZ A., AVRON M., "Production and selection of high beta-carotene mutants of *Dunaliella bardawil* (Chlorophyta)", *Journal of Phycology*, vol. 27, pp. 652–656, 1991.

[SHE 98] SHEEHAN J., DUNAHAY T., BENEMANN J., *et al.*, A look back at the U.S. Department of Energy's Aquatic Species program: biodiesel from algae, National Renewable Energy Laboratory, Internal Report, 1998.

[SHI 99] SHI X.M., CHEN F., "Production and rapid extraction of lutein and the other lipid-soluble pigments from *Chlorella protothecoides* grown under heterotrophic and mixotrophic conditions", *Food Research*, vol. 43, pp. 109–113, 1999.

[SHI 00] SHI X.M., ZHANG X.W., CHEN F., "Heterotrophic production of biomass and lutein by *Chlorella protothecoides* on various nitrogen sources", *Enzyme and Microbial Technology*, vol. 27, pp. 312–318, 2000.

[SOL 10] SOLOMON C.M., COLLIER J.L., MINE BERG G., *et al.*, "Role of urea in microbial metabolism in aquatic systems: a biochemical and molecular review", *Aquatic Microbial Ecology*, vol. 59, pp. 67–88, 2010.

[SOU 12] SOUZA P.O., FERREIRA L.R., PIRES N.R., *et al.*, "Algae of economic importance that accumulate cadmium and lead: a review", *Revista Brasileira de Farmacognosia*, vol. 22, pp. 825–837, 2012.

[SPO 06] SPOLAORE P., JOANNIS-CASSAN C., DURAN E., *et al.*, "Commercial applications of microalgae", *Journal of Bioscience and Bioengineering*, vol. 101, pp. 87–96, 2006.

[STA 98] STADNICHUK I.N., RAKHIMBERDIEVA M.G., BOLYCHEVTSEVA Y.V., *et al.*, "Inhibition by glucose of chlorophyll a and phycocyanobilin biosynthesis in the unicellular red alga *Galdieria partita* at the stage of coproporphyrinogen III formation", *Plant Science*, vol. 136, pp. 11–23, 1998.

[TAC 08] TACON A.G.J., METIAN M., "Global overview on the use of fish meal and fish oil in industrially compounded aquafeeds: trends and future prospects", *Aquaculture*, vol. 285, pp. 146–158, 2008.

[TES 10] TESTER M., LANGRIDGE P., "Breeding technologies to increase crop production in a changing world", *Science*, vol. 327, pp. 818–822, 2010.

[TRA 13] TRAN M., VAN C., BARRERA D.J., *et al.*, "Production of unique immunotoxin cancer therapeutics in algal chloroplasts", *Proceedings of the National Academy of Sciences*, vol. 110, pp. 15–22, 2013.

[TRE 98] TREDICI M.R., ZITTELLI G.C., "Efficiency of sunlight utilization: tubular versus flat photobioreactors", *Biotechnology Bioengineering*, vol. 57, pp. 187–197, 1998.

[TSY 01] TSYGANKOV A.A., "Laboratory scale photobioreactors", *Applied Biochemistry and Microbiology*, vol. 37, pp. 333–341, 2001.

[TUL 12] TULLI F., CHINI ZITTELLI G., GIORGI G., *et al.*, "Effect of the inclusion of dried *Tetraselmis suecica* on growth, feed utilization, and fillet composition of European sea bass juveniles fed organic diets", *Journal of Aquatic Food Product Technology*, vol. 21, pp. 188–197, 2012.

[UKE 76] UKELES R., ROSE W.E., "Observations on organic carbon utilization by photosynthetic marine microalgae", *Marine Biology*, vol. 37, pp. 11–28, 1976.

[ULU 11] ULUKAN H., "The use of plant genetic resources and biodiversity in classical plant breeding", *Acta Agriculturae Scandinavica Section B – Soil and Plant Science*, vol. 61, pp. 97–104, 2011.

[VAR 09] VARSHNEY R.K., NAYAK S.N., MAY G.D., *et al.*, "Next-generation sequencing technologies and their implications for crop genetics and breeding", *Trends in Biotechnology*, vol. 27, pp. 522–530, 2009.

[VIG 12] VIGEOLAS H., DUBY F., KAYMAK E., *et al.*, "Isolation and partial characterization of mutants with elevated lipid content in *Chlorella sorokiniana* and *Scenedesmus obliquus*", *Journal of Biotechnology*, vol. 162, pp. 3–12, 2012.

[VIL 11] VÍLCHEZ C., FORJÁN E., CUARESMA M., *et al.*, "Marine carotenoids: biological functions and commercial applications", *Marine Drugs*, vol. 9, pp. 319–333, 2011.

[VON 55] VON LIEBIG J.F., *Principles of Agricultural Chemistry: With Special Reference to the Late Researches Made in England*, Walton & Maberly, London, 1855.

[WAN 12] WAN L., HAN J., SANG M., *et al.*, "De novo transcriptomic analysis of an oleaginous microalga: pathway description and gene discovery for production of next-generation biofuels", *PLoS One*, vol. 7, p. e35142, 2012.

[WAN 08] WANG Y., PENG J., "Growth-associated biosynthesis of astaxanthin in heterotrophic *Chlorella zofingiensis* (Chlorophyta)", *World Journal of Microbiology and Biotechnology*, vol. 24, pp. 1915–1922, 2008.

[WIJ 10] WIJFFELS R.H., BARBOSA M.J., EPPINK M.H.M., "Microalgae for the production of bulk chemicals and biofuels", *Biofuels Bioproducts and Biorefining*, vol. 4, pp. 287–295, 2010.

[WIN 11] WINTER J.M., BEHNKEN S., HERTWECK C., "Genomics-inspired discovery of natural products", *Current Opinion in Chemical Biology*, vol. 15, pp. 22–31, 2011.

[XU 08] XU Y., CROUCH J.H., "Marker-assisted selection in plant breeding: from publications to practice", *Crop Science*, vol. 48, pp. 391–407, 2008.

[YAN 00] YANG C., HUA Q., SHIMIZU K., "Energetics and carbon metabolism during growth of microalgal cells under photoautotrophic, mixotrophic and cyclic light-autotrophic/dark-heterotrophic conditions", *Biochemical Engineering Journal*, vol. 6, pp. 87–102, 2000.

[YEH 09] YEH N., CHUNG J.P., "High-brightness LEDs – energy efficient lighting sources and their potential in indoor plant cultivation", *Renewable and Sustainable Energy Reviews*, vol. 13, pp. 2175–2180, 2009.

[YU 11] YU S., LIU S., LI C., *et al.*, "Submesoscale characteristics and transcription of a fatty acid elongase gene from a freshwater green microalgae, *Myrmecia incisa* Reisigl", *Chinese Journal of Oceanology and Limnology*, vol. 29, pp. 87–95, 2011.

[ZHA 13] ZHANG Y., WHITE M.A., COLOSI L.M., "Environmental and economic assessment of integrated systems for dairy manure treatment coupled with algae bioenergy production", *Bioresource Technology*, vol. 130, pp. 486–494, 2013.

[ZHA 11] ZHAO R., CAO Y.U., XU H., *et al.*, "Analysis of Expressed Sequence Tags From The Green Alga *Dunaliella Salina* (Chlorophyta)1", *Journal of Phycology*, vol. 47, pp. 1454–1460, 2011.

[ZHU 97] ZHU C., LEE Y., CHAO T., "Effects of temperature and growth phase on lipid and biochemical composition of *Isochrysis galbana* TK1", *Journal of Applied Phycology*, vol. 9, pp. 451–457, 1997.

[ZUC 01] ZUCCHI M.R., NECCHI O., "Effects of temperature, irradiance and photoperiod on growth and pigment content in some freshwater red algae in culture", *Phycological Research*, vol. 49, pp. 103–114, 2001.

Figure 3.17. *Example of microbial mats (French Polynesia)*

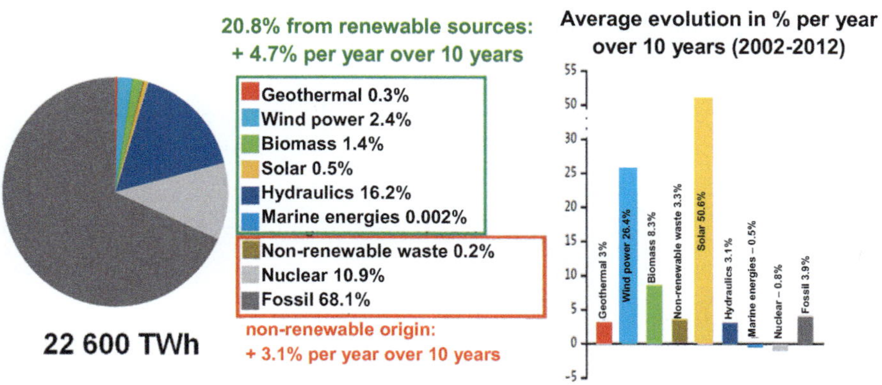

Figure 4.4. *Global energy production in 2012 and average rates, over ten years, of growth for each source [OBS 13]*

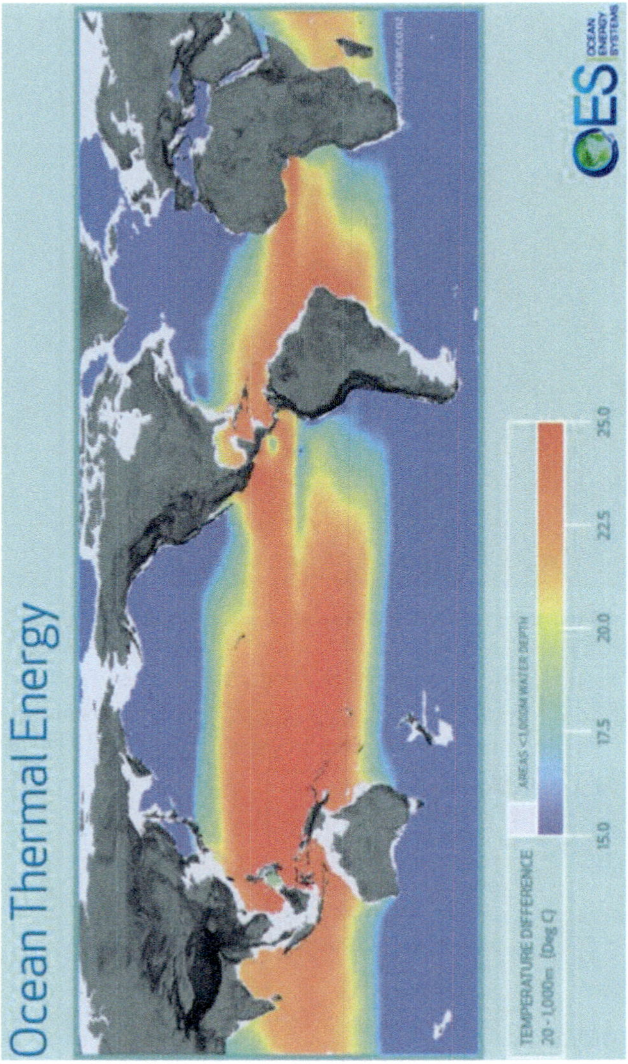

Figure 4.8. *Global OTEC resource [OES 12]*

Figure 4.16. *Punctual absorber*

Figure 4.17. *Oscillating water column*

Figure 4.18. *"Horizontal" or "vertical" submerged oscillating converter*

Figure 4.20. *OTEC demonstration floating module for Martinique (source DCNS)*

3

Pharmacology of Reef Marine Organisms

3.1. Introduction

The objective of this chapter is not to make an inventory of the molecules taken from reefs, nor is it a review of the vast literature on the subject which already exists [KOR 05]. The characteristics of the molecules taken from the marine environment and their reinforced pharmaco-chemical properties will be shown there. The experimental steps and methods used as well as their evolution from the organism collection in its environment to the description of molecules, their properties and exploitation are overviewed.

3.1.1. *Geographical strengths*

Seas and oceans cover 70% of the surface of the Earth, but are only home to 15% of the known biodiversity, with 230,000 described species. The main characteristic of aquatic populations is their large animal diversity: of the 33 known phyla, 15 are exclusively marine, such as the cnidarians and the echinoderms. The knowledge available on the different organisms varies according to the taxa: only a small number of species living in marine environments are described. A great part of the research is carried out in tropical or temperate areas,

---

Chapter written by Cécile DEBITUS and Jean GUÉZENNEC.

despite the fact that these regions present large biomasses; only around 3% of studies have been on organisms of polar regions, where they also have to adapt to extreme environmental temperatures, and fight against freezing.

While coral reefs represent less than 1% of the surface of the Earth, they make up more than a third of the marine species described. France, with its overseas territories and departments, is the fourth country with the fourth largest coral reef surface in the world behind the Philippines, Indonesia and Australia.

Marine biodiversity is concentrated in Indo-Pacific, Caribbean or Mediterranean hot-spots, and becomes poorer toward the East in the Pacific and Atlantic Oceans (Figure 3.1).

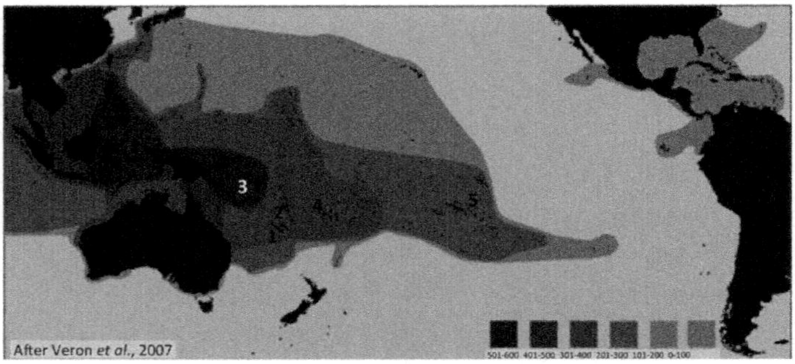

Figure 3.1. *Example of dispersion gradient: dispersion of hermatypic scleractinians*[1] *in the Pacific from the Indo-Malaysian center (1: New Caledonia, 2: Vanuatu; 3: Solomon Islands; 4: Fiji; 5: French Polynesia) [VER 09]*

New Caledonia and French Polynesia are two French overseas territories (OTs) presenting important qualities for the study of the natural resources of their reefs: New Caledonia, with its proximity to the Indo-Pacific biodiversity center, hosts the largest barrier reefs and lagoons in the world and French Polynesia, covering a huge geographical area, is home to a large diversity of coral formations, including 20% of the world's atolls. These archipelagos are the places of choice for the development of bioprospection programs.

1 Corals that build reefs while living symbiotically with microalgae.

3.1.2. *The marine environment: a source of new molecules*

3.1.2.1. *The development of marine chemistry*

Marine chemistry only really started in the 1970s, with the development of scuba diving, allowing scientists to access marine biodiversity, especially benthic organisms. The marine environment is of considerable interest for research, especially in the field of natural products of therapeutic or pharmacological interests.

The evolution of the literature (Figure 3.2) shows not only a slight slowing down of the number of new natural marine molecules discovered each year, but also of synthetic and semi-synthetic methods[2] in the study of marine molecules. These methods are used to confirm complex structures of natural compounds, to build simplified analogs or to obtain enough active pure compounds for further biological activity studies.

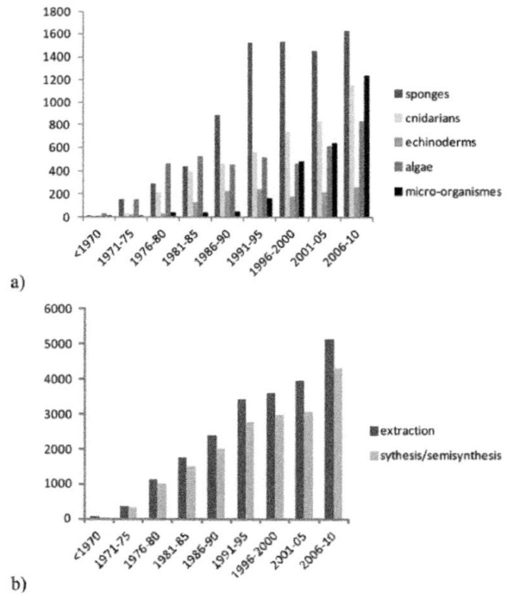

Figure 3.2. *Evolution of the number of molecules published in the literature: a) in the main groups studied and b) comparison extractive chemistry and synthesis (graph produced with data from the MarinLit database, University of Canterbury, 2012 version)*

2 Chemical transformation of an existing natural molecule.

The complexity of these marine natural products often offers quite an intellectual challenge for organic chemists. Currently, approximately 18,000 molecules are described, with around one-third coming from sponges.

Sponges are present in all marine environments and represent the second phylum living on coral reefs in terms of biomass. Their abundance and diversity facilitate the pharmaco-chemical studies that are difficult in rarer or smaller species. Algae are also abundant in temperate zones as well in tropical areas, but represent less than 20% of the sources of new molecules. Research on microorganisms, which are of special interest in biotechnology, is ever-increasing: their importance is obvious in the elaboration of new products in a huge array of application fields, from nutrition to medicine and also the environment (see section 3.4).

However, the cost of research and production of marine molecules orientates their development toward products with high added value, such as in healthcare and cosmetics.

3.1.2.2. *Chemical characteristics of marine molecules*

Marine molecules come from biosynthetic pathways that lead to the various "classical" chemical families: terpenes[3], polyketides[4], amino acids and derivatives (alkaloids, peptides and nucleotide bases). These different chemical families are present in every studied group (Figure 3.3). The terpenes, which play a major role in the structure or physiology of cells, as well as in chemical communication, dominate in all the groups studied.

These molecules are very often halogenated or sulfated: bromine is the most frequent halogen as much in phenolic compounds as in terpenes or alkaloids. It is present at a lower concentration than chlorine in the marine environment (around 8 mmol), but its incorporation is linked to the existence of haloperoxidases, like the bromoperoxidases highlighted in red algae. Most of their brominated

3 Class of hydrocarbons produced by plants, metazoan and aquatic organisms, such as cnidarians or sponges.

4 Secondary metabolites of bacteria, mushrooms, plants and animals.

molecules are terpenes. Some compounds may bear several different halogens (Figure 3.4).

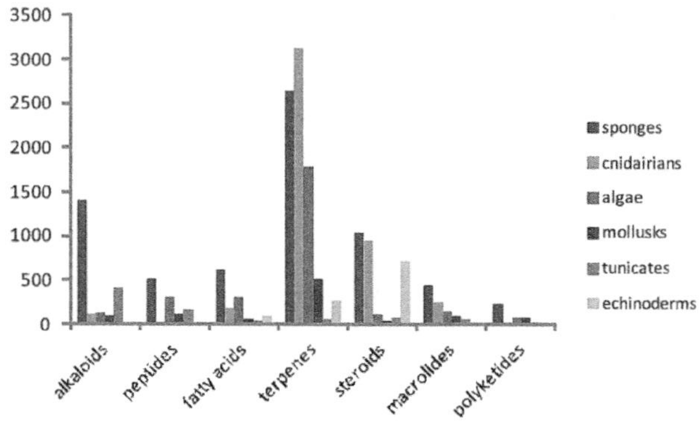

Figure 3.3. *Distribution per type of metabolite in the main groups studied (graph produced with data from the MarinLit database, University of Canterbury, 2012 version)*

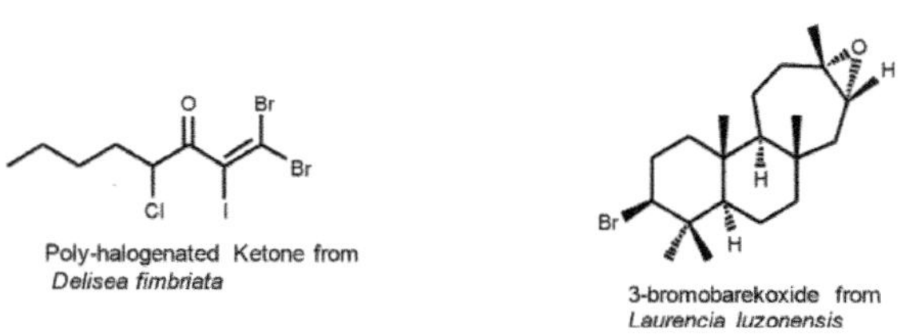

Figure 3.4. *Examples of halogenated algae terpenes*

In animals, sponges present the highest content of brominated molecules, especially some chemotaxonomic alkaloid markers,[5] such as the tyrosine derivatives in Verongida, or aminoimidazoles in the Agelasida (Figure 3.5).

Sulfur appears as much in primary as in secondary metabolites, in marine plants as in invertebrates. It is thus incorporated in different forms in a variety of compounds (Figure 3.6): thio-ethers, isothiocyanates, sulfates, sulfones, disulfure, etc. In sponges, taurine is commonly grafted to alkaloids or terpenes skeletons.

5 Chemotaxonomy: classification of organisms using secondary metabolite structures.

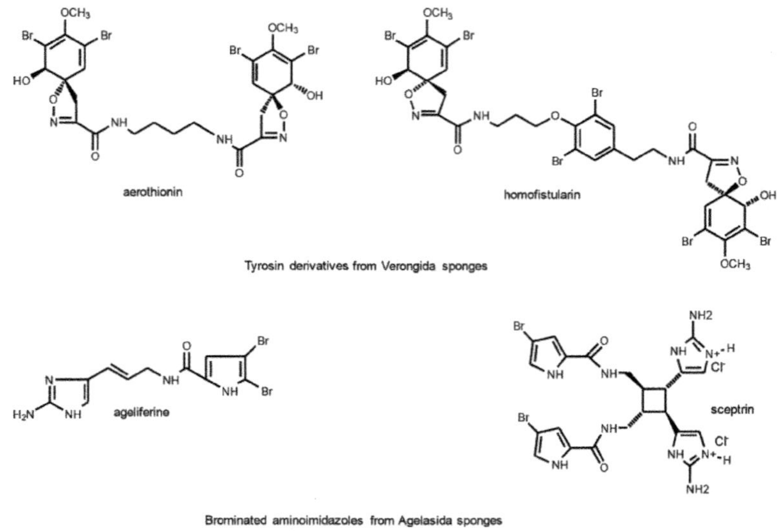

Figure 3.5. *Examples of characteristic brominated alkaloids of sponges*

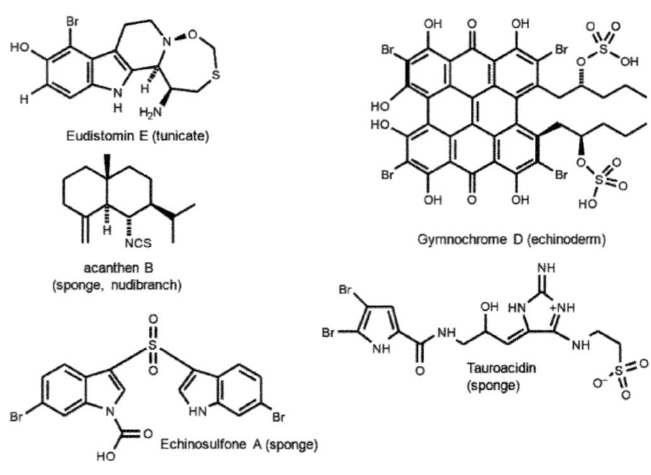

Figure 3.6. *Examples of sulfated marine molecules*

Some rare elements are encountered in marine organisms: arsenic is incorporated in some examples of brown algae and their predatory mollusks, as well in fish and in one spiny lobster where these compounds are probably of commensal origin (Figure 3.7). More recently, a compound containing several atoms of arsenic, arsenicin, was found in a sponge discovered off the New Caledonian coast. This molecule presents remarkable antimicrobial activity [MAN 06].

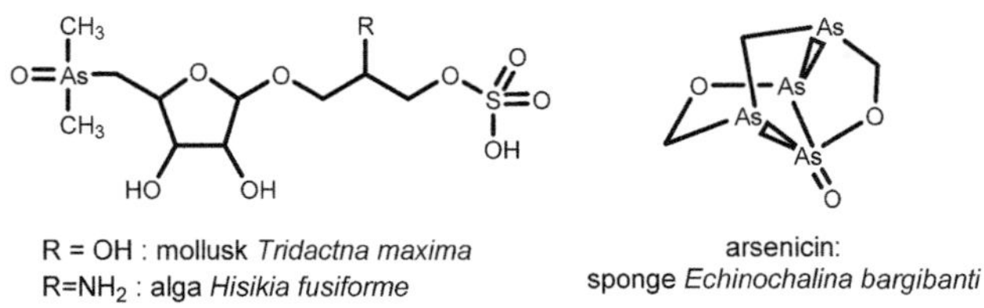

Figure 3.7. *Example of marine arsenic-containing molecules*

A good number of molecules have been isolated as metal complexes. Porphyrins complexing a nickel atom have been isolated from ascidians [RIN 88], while molecules of the same skeleton remain free in sponges [DAM 89, DAM 93]. Remarkable examples (Figure 3.8) include imidazolic alkaloids isolated from calcareous sponges of the *Leucetta* genus in the form of zinc complexes [MAN 95]; these compounds are also present in nudibranch mollusks that feed on them.

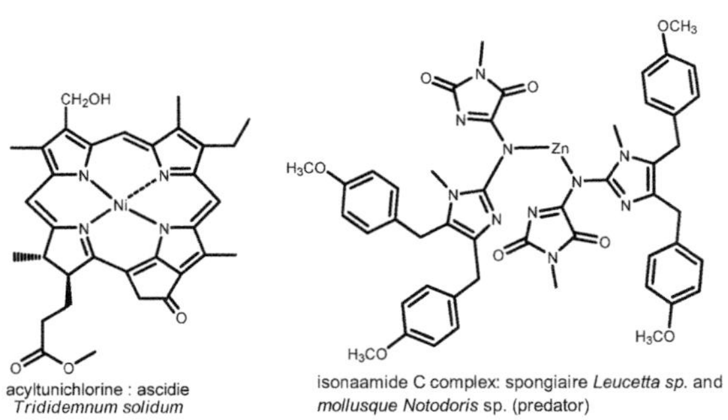

Figure 3.8. *Examples of metal complexing marine molecules*

Other examples of metal complexing have been described, in particular in depsipeptides (antimicrobial hormones), but the physiological role of these complexes in their host-organism remains unknown to date.

3.2. Applications of marine molecules: reality

3.2.1. *Acquired results*

The literature is abundant on the subject, and is summarized here. The potential of these molecules covers various domains: molecular models for the investigation of biological mechanisms or the elaboration of compounds for use as healthcare products or for environmental applications. Knowledge of the marine environment is relatively recent as the development of *in situ* exploration of the submarine world by humans only started in the 1940s, while human interaction with the terrestrial and plant world has been going on for over 3,000 years. Unlike for terrestrial organisms, which represent about 60% of the pharmacopeia, there is no ethno-pharmacological guide for the research of bioactive substances in marine organisms: they are rarely used traditionally (medicine, poison, etc.). Algae have long been used by humans as food, or for their hemostatic properties, against gastro-esophageal reflux (alginic acid and alginates) or hyperphosphoremia (carrageenans). Many cosmetic products incorporate algae extracts, and are particularly in vogue in balneotherapy and thalassotherapy.

More recently, in the last 30 years, the potential of the secondary metabolites of algae, invertebrates and, more recently, of marine microorganisms has continued to increase. More than 15% of substances of marine origin described in the literature present a level of biological activity in the different domains of human health: infections, cancer, inflammation, cardiovascular or nervous system, etc. (Figure 3.9). The future of marine molecules is orientated toward human health, with high added value products, but only a small number of these substances is currently – or will be – exploited pharmacologically, as the molecule must satisfy many criteria: activity, stability, lack of toxicity, as well as availability.

While the role of marine secondary metabolites for marine organisms is rarely known, these compounds do often possess original and active structures on different pharmacological targets. Apart from their true quality as “chemical weapons”, secondary metabolites taken from the marine environment are the subject of great interest for their applications for human use, for which they are exploited via the production of these high added value molecules.

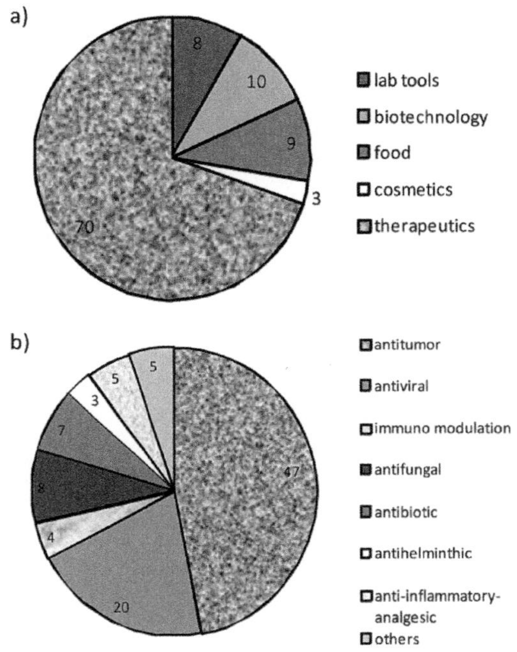

Figure 3.9. *Distribution of patents (%) a) in the different domains of utilization and b) in the different therapeutic domains [BON 96]*[6]

The discovery of spongothymidine and spongouridine (Figure 3.10) in the 1950s was a major starter to the rise of "marine chemistry": these compounds, isolated from a Caribbean sponge, *Cryptotethya crypta*, demonstrated antitumoral activity, and analogs have been elaborated as antitumoral (cytarabine: treatment of lymphoma and leukemia) and antiviral (vidarabine: treatment of chronic hepatitis C) medication. These quite old molecules have been "modernized" by fluorination especially to facilitate administration, as well as to improve their activity. AZT, the first major AIDS medication, is also a synthetic derivative of cytarabine. Vidarabine,

6 These numbers have not changed much since their publication.

known as a synthetic molecule, was discovered in 1984 as a natural product in the Mediterranean gorgonian *Eunicella cavolini*, and in this example, nature catches up[7] with human technology.

Two medicines have recently been released onto the market: an analgesic, Ziconotide®, which is a peptide isolated from the venom of the conidae *Conus magus*, and Yondelis (or Trabectedin), and antitumoral isoquinoleic alkaloid (ET743) identified from an ascidian *Ecteinascidia turbinata* (Figure 3.11). Ziconotide® (or Prialt®) is a powerful analgesic used to replace morphine when it is no longer effective in patients. Conidae venoms are the subject of a work package of the ambitious research program "Venomics", which has the double goal of providing new molecules and new tools for studying the physiology of different domains.

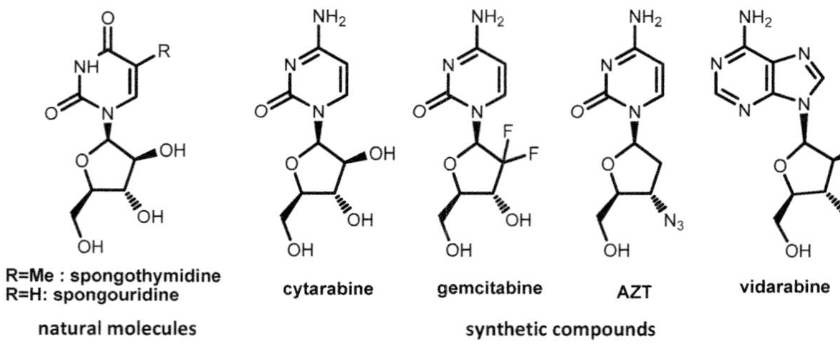

Figure 3.10. *Commercialized antitumoral and antiviral molecules derived from molecules of sponges*

Presently, Yondelis has a status of orphan medication and presents special activity against leukemia, melanoma, as well as against solid tumors (breast, ovary and lung). It is currently obtained by extraction from animals raised in a natural environment (mangroves). Several alternative production pathways for this molecule are under study, as reliable and economically worthwhile production is an essential step for the development of a new product.

7 The synthesis of the molecule was achieved before the discovery of the natural molecule.

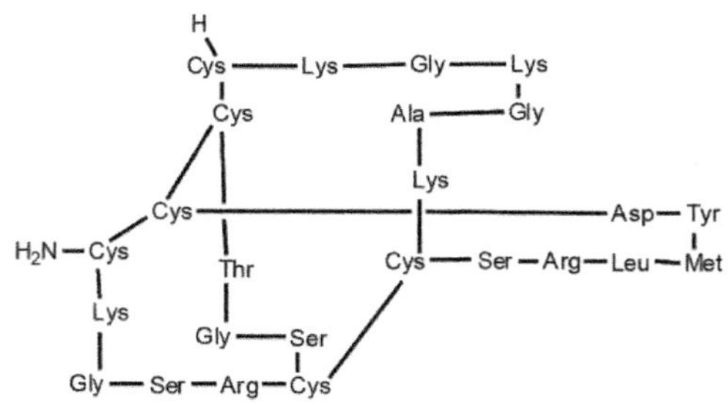

Ziconotide® or Prialt® (conotoxin SNX111)

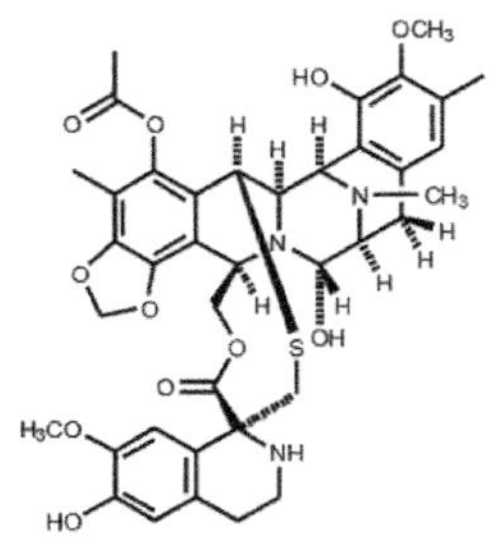

Yondelis®

Figure 3.11. *New compounds from marine invertebrates released onto the pharmaceutical market*

Marine animals are more discrete sources of compounds for use in cosmetics: an anti-inflammatory molecule, pseudopterogorgine from the Caribbean gorgonian, *Pseudopterogorgia elisabethae*, has been developed as the bioactive compound of a line of cosmetic products.

3.2.2. *Current trends*

The number of targets indentified has considerably increased since the 1990s, thanks to the large sequencing programs concerning the human genome and, for example, *Plasmodium falciparum*. These targets allow better and quicker diagnostics, and also enable us to discover new treatments. Many key proteins of these pathogens have been developed as targets for the search of new medications. The development of high-throughput screening based on molecular targets has led to the development of large molecule libraries to fulfill the needs of these high-capacity techniques. Combination chemistry, initially designed for the synthesis of collections of peptides of oligonucleotides, has become orientated toward the creation of libraries of small "drug-like" molecules[8]. Natural products were left aside for a while in favor of these large number of techniques. This direction has not fulfilled the hope for the development of a large number of *leads*[9], since the number of new molecules developed has dropped sharply since the 1990s.

The current trend is to favor the quality of the molecules tested over quantity. Combination chemistry on models of natural products is being developed, favoring the structural diversity of collections of screened molecules. Natural molecules have been refined over the course of evolution to reach their target: the biomimetic synthesis of natural compounds provides a variety of simplified intermediaries, likely to carry out the activity of the final molecules. The best molecules are often the least common, the most complex, and their availability for different studies, from the laboratory to the clinic, is vital. The current trend is the synthesis of analogs allowing some flexibility in the modification and functionalization of synthetic intermediates.

The example of bryostatin 1 illustrates the conception of a molecule from a natural model. This molecule and its natural analogs (18 compounds in total) isolated from a bryozoa, *Bugula neritina*, from the California Gulf, interact with protein-kinase C. This compound is present in trace amounts, in the order of 1 mg in 1 ton of organisms; even if the bryozoan grows easily on various supports as a "fouling" organism, the yield in terms of bioactive molecule is not enough to warrant its further development from the natural source.

The microorganism associated with the bryozoan that produces this compound has been isolated, but production using biotechnology is also difficult and not very satisfactory. However, the activity of this compound has been convincing enough to take it to clinical trials,

8 Drug-like molecule: molecule of molecular weight and complexity that is compatible with pharmaceutical development.

9 *Lead* = molecule at the head of a new therapeutic chemical series.

currently at phase 2. At the same time, synthetic analogs have been produced, but often also too complex for further development. The key elements of the activity of this molecule have been identified, in particular the position of the oxygen atoms of the macrocycle. They are required to maintain interactions with the target protein. The rest of the molecule has been simplified to get to analog A [PAT 05] (Figure 3.12), which is even more active than the natural molecules and which can be produced on a large scale: roughly 20 molecules of marine origin are currently in phase 2 or 3 of clinical trials. These molecules often have a complex structure with several functionalities and asymmetrical carbons, but this complexity is no longer an obstacle for their development. The analysis of the molecule and active sites of its therapeutic target, as well as the modeling work, makes possible the synthesis of bioactive analogs with a good enough yield.

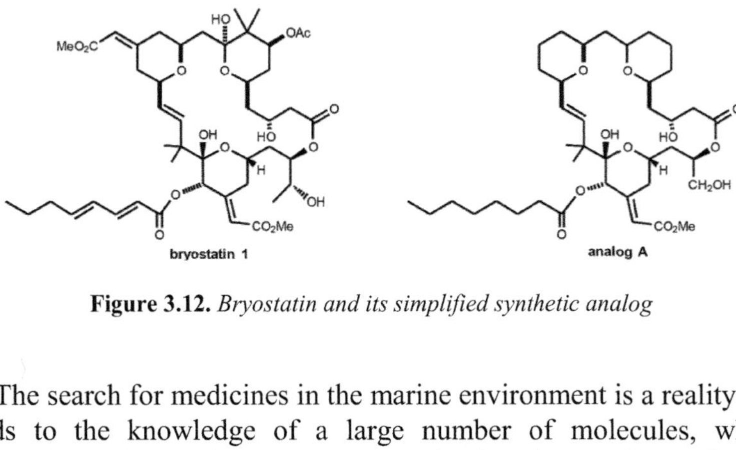

bryostatin 1 analog A

Figure 3.12. *Bryostatin and its simplified synthetic analog*

The search for medicines in the marine environment is a reality and leads to the knowledge of a large number of molecules, which represent a chemical treasure chest, despite the small number of molecules developed in this domain.

3.3. Concrete examples of marine natural products research programs

For over 30 years, the marine chemistry teams of IRD (formerly ORSTOM) have led programs to investigate marine resources in the tropical Pacific, a zone that is particularly favored for such research. The study zone lies in the South Pacific from the Western Pacific

(New Caledonia, Vanuatu and Solomon Islands) to the Central Pacific (French Polynesia), passing through the islands of Fiji (Figure 3.1). The archipelagos of Melanesia are located close to the golden Indo-Malaysia triangle, and present a large diversity of benthic invertebrates, at all taxonomic levels; moreover, the Norfolk ridge, presenting seamounts between 300 and 500 m deep, has been the home of the discovery of a large number of "living fossils" and their secondary metabolites of very special interest.

French Polynesia, in the center of the Southern Pacific, represents half of the French EEZ: it is as large as Europe over a great range of latitudes, between 8 and 27° S. Its islands and reefs are mainly isolated, leading to speciation and a particularly high rate of endemism for some benthic groups, such as mollusks (20%). Similar collection and research efforts on sponges have led to three or four times fewer species in French Polynesia than in the Solomon Islands, for example, but 30% of new species for the Pacific region, maybe even endemic, for only 4% in the Solomon Islands [HAL 13].

3.3.1. *Steps of a project*

3.3.1.1. *Access to biodiversity and benefits sharing*

A guide on this subject for use by researchers has recently been edited [ANO 11]. Wherever the sites of the collections are, these are the object of mandatory authorizations (MOU for Memorandum Of Understanding) from the authorities (governments and local provinces, fishing services, environment, education and research offices) and local communities (communes and landowners) resulting in a permit for the collection, research and export of the organisms collected.

Benefit sharing, besides a random monetary return in the very long term for the case of the commercialization of molecule, is represented by efforts to train young researchers and technicians from the countries of origins of the organisms, as well as returning the results to those countries. More recently, the access of partners, decision-makers and stakeholders to the online database Cantharella[10] makes it possible to follow the evolution of the research carried out on the collected organisms in their countries.

3.3.1.2. *From sampling to the molecule: evolution in favor of the preservation of the environment*

Collections in the 0–60 m zone are carried using scuba diving, allowing the observation of organisms in their environment, selective collection as well as *in situ* photography, which is very useful in the identification of these often shapeless invertebrates, unrecognizable once taken back up to the surface. Warren dredges or beam trawl, depending on the nature and slope of the sea bottom, are used to collect bathyal benthic organisms (Figure 3.13). The organisms are brought up of course unsorted, and with a few exceptions have very different appearances from what they have *in situ*.

Figure 3.13. *a) Research Vessel Alis, b) beam trawl, c) Warren dredge and d) contents of the dredge to be sorted (photos: IRD/Bertrand Bourgeois a), Bertrand Richer de Forges b) and c), and Cécile Debitus d))*

However, careful washing and sorting allows us to obtain sets of samples of the different organisms for identification and chemical

10 http://cantharella.ird.nc.

study. These methods have proved to be reproducible each time that it has been necessary to gather more of the organisms of interest.

The reference samples from these prospection campaigns are kept in international museums to ensure their sustainable conservation and access for the scientific community, and are identified by specialists of each group.

The methods of extraction, analysis and pharmacology have evolved toward miniaturization, allowing lower sampling from several kilograms to a few grams. This evolution not only allows access to a greater biodiversity (less abundant or small species), but also helps in environment conservation. The use of an accelerated solvent extraction (ASE) technique makes possible the extraction of smaller samples, thus preserving the natural resource and increasing the range of biodiversity to investigate, as well as reducing the amount of organic solvents, which can have negative effects on the environment, and creating collections of standardized samples for a better valorization.

The methods of chromatography and structural analysis have also experienced great improvement in performance, and can now deal with complex structures with only a few micrograms.

The search for biological activities has also been transformed by the evolution of molecular and cellular biology and the development of high-throughput screenings: old hand-made screening on isolated organs or cell cultures required several milligrams of extract, while the bioactivity on enzymatic targets can now be worked out with only a few micrograms, allowing the extract to be kept for the isolation of the active molecule(s).

While the evolution of identification methods and characterization of the bioactivity of the molecules allow minute quantities to be used, organic synthesis also had to adapt to allow the development of these compounds. Biomimetic synthesis[11] in many cases can result in excellent yields, compatible with production on a scale allowing

11 Biomimetic synthesis: chemical synthesis that imitates the biosynthetic pathways of natural molecules.

further clinical trials, development and commercialization of relevant compounds.

3.3.1.3. *Outstanding results*

The marine programs carried out by the IRD and its partners in the Pacific have resulted in over 200 publications, mainly on new molecules, most of which are active on one of the biological models studied. Two outstanding examples of applied or academic results are described below.

Girolline, or girodazole (RP49532A), is an important result from these Pacific research programs. This molecule specifically inhibits the synthesis of proteins with an original mode of action at the level of the end of the synthesis of proteins: it blocks the release of the peptides out of ribosomes [COL 92]. This small molecule (Figure 3.14), with remarkable antitumoral properties, reached phase 2 of clinical trials in humans using the natural product (*Cymbastela cantharella* is a fast growing sponge found abundantly in the New Caledonian lagoon), but was abandoned due to excessive side effects (hypotensor).

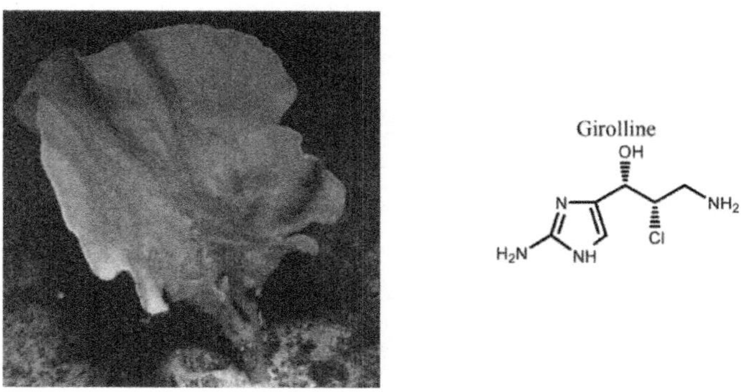

Figure 3.14. *Cymbastela cantharella., source of Girolline (photo: IRD/Jean-Louis Menou)*

A partnership with the CNRS and the pharmaceutical industry led to the synthesis of many derivatives with the aim of reducing these side effects, but without success: the natural product remains the most active. This product, for which asymmetrical synthesis has been developed,[12] is always evaluated for different pharmacological targets. It presents an original mechanism of inhibition of protein synthesis in the domain of oncology.

The investigation of *Gymnocrinus richeri* was first carried out for its special taxonomic interest, as soon as this exceptional creature was discovered on submarine mounts at about 500 m deep. This led to the discovery of new molecules (Figure 3.15), the gymnochromes [DER 91]: the skeleton was known, as hypericin-like, but also in fossil crinoids from Swiss quarries (the fringelites) [BLU 62], and very recently analogs bearing lateral non-brominated chains were found in fossilized crinoids of Polish quarries [WOL 08]. A new example was described recently in another pedunculated crinoid of the bathyal zone [WAN 10], which reinforces the interest in terms of chemotaxonomy of these molecules in crinoids.

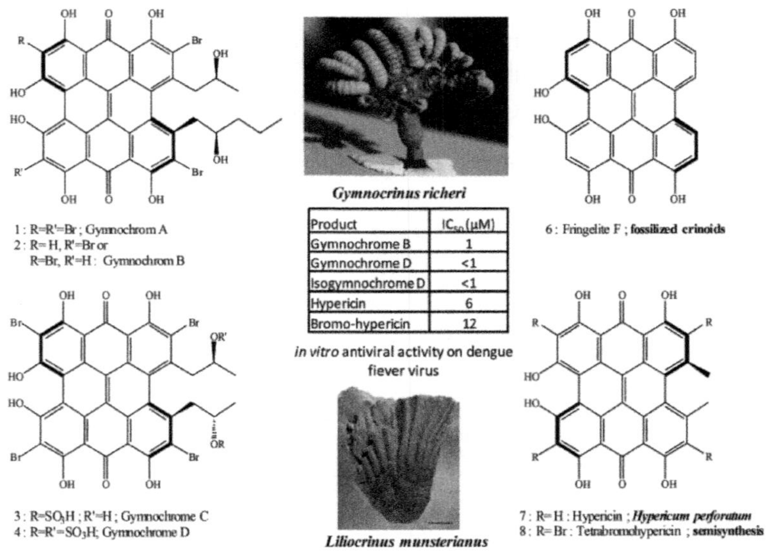

Product	IC_{50} (µM)
Gymnochrome B	1
Gymnochrome D	<1
Isogymnochrome D	<1
Hypericin	6
Bromo-hypericin	12

Figure 3.15. *Structure of the gymnochromes of Gymnocrinus richeri and their fossil and plant analogs, activities provided in the dengue virus, in vitro (we would like to thank Dr Wolkenstein for providing us with the photo of Liliocrinus munsterianus. G. richeri: photo IRD/Pierre Laboute)*

---

12 Asymmetrical synthesis: synthesis respecting the stereochemistry of the natural molecule, also called "stereoselective".

A highly original precursor work was carried out in collaboration with the *Institut Pasteur de Nouvelle-Calédonie* to find out active substances against dengue fever virus, a viral disease transmitted by mosquitos. The symptoms of dengue fever are similar to the flu [LAI 98]. First of all, we tested five molecules, for which we had previously determined the level of antiviral activity, either against HSV or HIV. Only gymnochrome B proved to be active, and we therefore evaluated the activity levels of minor compounds, as well as hypericin (commercial) and its tetrabrominated derivative [LAU 05]. These compounds have long remained the only molecules to inhibit *in vitro* replication of the dengue virus. The activity levels of gymnochromes are stabilized compared to those of hypericin, by bromide and their lateral chains. The sulfate groups of the lateral chains also significantly improve activity.

As a result of the important spread of dengue in different regions of the world, the search for medication against this virus has gained importance. Plants, particularly those used in traditional medicine, have given rise to screenings on the replication of the virus or of enzyme targets. Only flavonoid-type polyphenols (like gymnochromes) [KEI 11] have been identified as active ingredients for natural products, and synthetic analogs of nucleosides for synthetic molecules [TIC 12], these nucleosides having already shown promising characteristics in cancerology and virology (see section 3.2.1).

This academic work perfectly illustrates the notion of the different values of the natural resources: values of options (medication or potential laboratory tool) and cultural value. The invertebrates, like those studied in bioprospection programs, do not have any value of direct consumption nor cultural value, except for a few mollusks and some sea cucumbers used as food or medicine or also for fishing or jewelry.

3.4. Marine environment and biotechnology: the essential role of microorganisms

Aquaculture for anything other than food or cosmetics remains exceptional. Toiletry sponges were among the first sponges cultivated in history. Some attempts were made to cultivate some invertebrates *in vivo* for their secondary metabolites in clinical development, but this option has not been favored as a source of bioactive compound. Some *in vitro* cultures have been carried out for academic investigation of these organisms, or for the *in vitro* elaboration of spicules with electronic applications for their unique physico-chemical properties. Due to their large filtration capabilities, sponges have found their applications in bioremediation.

Works in biotechnology for the production of secondary metabolites and active ingredients are mainly focused around microorganisms and in particular bacteria. Since the 1930s, around 50,000 molecules have been discovered from microorganisms, more than 10,000 of which are biologically active, the majority in the domains of antibiotics and antitumorals. Around a hundred products of microbial origin are used today in human healthcare and agricultural chemistry. Special attention has been given to microflora of sea beds, especially actinomycetes, from which most secondary metabolites described originate. While the properties of these bacteria are still being explored, it is extremely difficult to discover new molecules: 90% of active cultures provide known molecules, which are costly and inefficient. The development of resistance, particularly in *Mycobacterium*, the bacterium responsible for tuberculosis, stimulates the exploration of new sources. The diversity of the oceans is linked to variations in temperature, pressure and salinity: marine microorganisms have developed unique capabilities for metabolic and physiological adaptation to survive in extreme environments, and thus offer a potential for the production of metabolites that would not be present in terrestrial microorganisms. Marine microbiology benefits from technical progress as much in genomics, molecular characterization or elaboration of pharmacological targets, and also for chemical and structural analysis. The large structural diversity of their secondary metabolites also reflects the biodiversity of these organisms and their origins.

3.4.1. *Microorganisms of the marine environment*

The marine world benefits from a wide diversity of environments, but has long suffered from the belief that its microorganisms are difficult to cultivate [SOL 04]. It is true that only a small number of microorganisms can be cultivated in standard environments, but isolation and culture of most of the known genera can be carried out on most of the common types. Most marine bacteria do require a high concentration of salts, causing some technical issues. Molecular systematics has recently resulted in the identification of the organisms forming the population of a sample and the adaptation, in relation to the knowledge of the specificities and the original environment, of the culture media so as to isolate the microorganisms of interest.

Seawater plankton bacteria are essentially gram-negative bacteria that are not particularly prolific as active secondary metabolites. Marine sediments offer an environment far richer in nutrients, whose composition varies with different biotopes: temperate or tropical shallow coastal waters, geological sediments of various chemical compositions, or even proximity with hydrothermal sources or low-oxygen environments.

The investigation of marine actinomycetes, which is less abundant than on land, started in the 1970s and has developed since successfully, with most active secondary metabolites of marine microorganisms now coming from this source. This development has greatly benefited from progress in genomics, as much for knowledge on these microorganisms, as for screening methods. No molecules are currently in clinical trials, but this should change over the next few years.

Benthic marine organisms are also host to a large number of microorganisms: microalgae, bacteria, fungal strains, etc. Mollusks owe their seasonal toxicity to these microorganisms, making them inedible. Some questions remain unanswered to this day: ascidians of the Didiemnidae family are associated with the prochlorophyceae (*Prochloron* sp.). A good number of metabolites seem to be produced by these microalgae, which are easy to isolate, but no cultures of *Prochloron* sp. have yet been successful: they cannot survive without their host, which provides elements that are not yet reproducible *in vitro*. Sponges can contain up to 50% of their weight in bacteria and, therefore, represent an interesting source of these microorganisms, knowing that a large proportion of these come from seawater itself.

The low content of active molecules in marine macroorganisms and their structural complexity often leads to a search for their origin in the associated and/or symbiotic microorganisms. *In vitro* controlled production of complex molecules means the natural resource is no longer required, and, unlike synthetic chemistry, produces few synthetic subproducts or coproducts. This has been attempted for several promising molecules for which the extraction yield is particularly low, and for which the chemical synthesis is complex and not worthwhile. A lot of work has been carried out on the origin of bryostatin: the polyketide synthases, responsible for the biosynthesis of bryostatin, have been isolated and identified in the symbiotic bacterium, and hypotheses on the biosynthesis have been proposed, but the molecule has not yet been obtained *in vivo*.

Sponges are organisms of choice for work in microbiology. A genomic approach by analysis of the sequences of 16S rDNA has made possible the analysis of the bacterial content, cultivatable or not, of sponges and has shown that these organisms are hosts to an important population of actinomycetes that were thought to be located largely in sediments. A recent example of success is described for a type of antimalarial alkaloid, the manzamines, first isolated from sponges. Several manzamines have been isolated from different genera of sponges, which would suggest that their origin lies in an associated microorganism. Manzamine A and its hydroxylated derivative (Figure 3.16) were obtained from cultures of *Micromonospora* sp., the primary resource being tropical sponges.

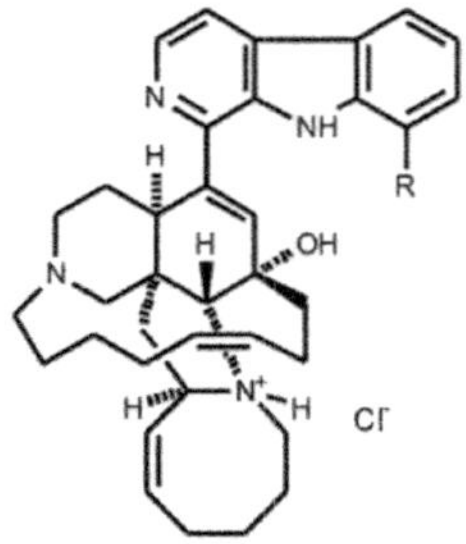

R=H : Manzamin A, Yield = 0.011%
R=OH: 8-hydroxymanzamin A, Yield = 0.003%

Figure 3.16. *Manzamine, antimalarial alkaloid obtained from the culture of an actinomycete*

The study of marine fungi only started developing over the last few years, despite the fact that one of the oldest and most important molecules in therapeutics, cephalosporin, comes from this source. They produce a great diversity of active secondary metabolites and have been the subject of several recent reviews [SAL 07]. They can be split into two groups: obligate or facultative halophytes. They can be isolated from all environments and from very varied substrates: water, sediments, invertebrates, sunken or living wood (mangroves), algae, etc., more than 70% coming from sponges, algae or wood. Trials have been carried out, without success to this day, to obtain *in vitro* the compounds identified in the total extracts of the host organisms through the culture of isolated fungal strains of these hosts. For example, sponges of the genus *Zyzzya* are rich in pyrroloiminoquinones: the makaluvamines; trials with fungal strains have been carried out without success to obtain these compounds *in vitro*, one example of which is, however, produced by a culture of "terrestrial" myxomycetes [ISH 01].

3.4.2. *Example of an atypical ecosystem: microbial mats, or kopara*

Marine ecosystems provide very special physical and chemical properties (high pressures, pH, temperature, redox potential gradients, variable salinity, high contents of organic and inorganic toxic elements, etc.) and are wonderful fields of investigation for the search for molecules of interest in biotechnology, for example bioactive metabolites. Marine hydrothermal sources, the Arctic and Antarctic continents [GUE 02, MAN 05] and microbial mats, such as those present on a few atolls of French Polynesia [DEF 94], are great fields for such research. These microbial mats, sometimes several centimeters thick and called *kopara* by the inhabitants of these atolls (see Figure 3.17), have for several years been the subject of investigations around the biotechnological potentials of the microorganisms [GUE 11].

These microbial mats usually host populations of cyanobacteria and aerobic bacteria on their surface, while anaerobic bacteria develop in the lower layers, due to their limited oxygenation and where sulfate-reducing bacteria can be found. The physical and chemical conditions, especially their variations linked to the geographical location of these microbial mats, may allow them to host populations of microorganisms that have developed strategies of adaptation to these atypical conditions and are therefore considered as potential sources of molecules of interest in biotechnology.

Figure 3.17. *Example of microbial mats (French Polynesia) (see color section)*

To date, a few studies have focused on the search for bioactive molecules or secondary metabolites from these microorganisms. However, preliminary studies have shown the presence of antimicrobial molecules, although precise identification of these molecules has not yet been started. However, an important pigment for ophthalmology, 3 R-R'zeaxanthine has been produced in laboratory conditions by a culture of *Paracoccus sp.* isolated from these microbial mats.

Along this search for bioactive molecules and new medication, bacterial biopolymers can also provide numerous opportunities of discoveries in the fields of new materials and/or new physical, chemical and biological properties. These biopolymers produced *in vitro* present some advantages, such as the absence of dependence with regard to meteorological, ecological and political changes that could affect the quality, cost and provision of similar products extracted from algae, plants or animals. Production can be optimized by acting on conditions of fermentation (sources of carbon, temperature, aeration, pH, etc.). The use of cultured bacteria to produce biopolymers ensures traceability, as well as allowing modifications of the product. Two types of biopolymer stand out in terms of biotechnological applications, exopolysaccharides (EPSs) and polyhydroxyalkanoates (PHAs).

3.4.2.1. *Exopolysaccharides*

Polysaccharides can be defined as macromolecules formed by the linking of similar motives, in this case carbohydrates, commonly called sugars or monosaccharides. Initially dominated by plant or animal rubbers, the market is also opening to bacterial polysaccharides. In bacteria, these polysaccharides are present at the level of the cell wall as an essential element of lipopolysaccharides (LPSs), outside the cell, where they are either linked to the same cell (capsular polysaccharide or CPS) and surrounding the bacterial cell, or even back in the culture medium in the form of EPSs.

In nature, these EPSs play an important role in the survival of bacteria by influencing their physical and chemical environment. They fulfill several functions such as a physical barrier, favoring the fight against phagocytosis or even against desiccation, and ensure an effective protection against low and high temperatures, or even salinity. They are, therefore, a response to environmental stress, and also a protection against predators and the presence of toxic organic elements. Finally, they can play an essential role in the processes of cellular interaction and recognition, in particular in the irreversible fixing of microorganisms on exposed surfaces in the natural environment. Indeed, once fixed on a support, the microorganisms trap organic matter and the dietary elements necessary to their growth by forming a biological film, or biofilm. This biofilm is often the first step of sequences that lead to the formation of micro- and macrofouling, with all its consequences, as much on the level of heat transfer, resisting fluid flow, or even at the level of the initiation of the most well-known phenomena of corrosion, *microbially induced corrosion* (MIC) (Figures 3.18 and 3.19).

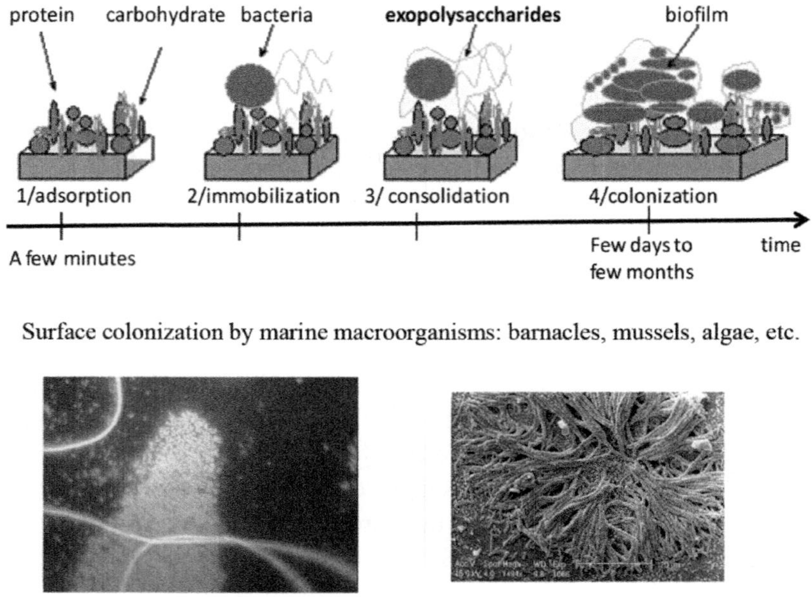

Figure 3.18. *Biofilm formation by exopolysaccharide production*

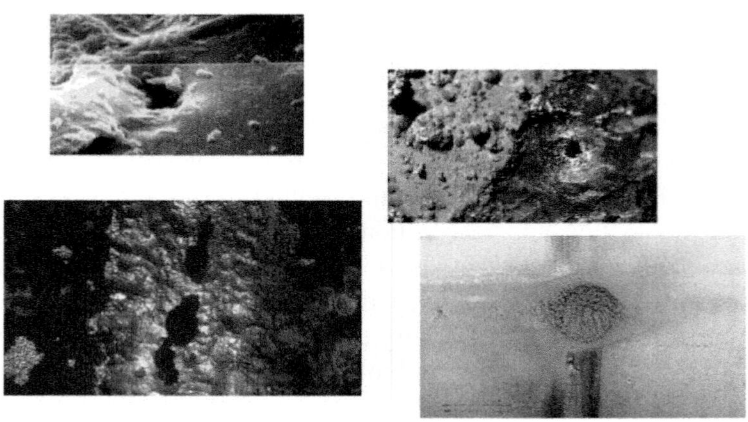

Figure 3.19. *Examples of corrosion induced by the presence of microorganisms and biofilms*

Polysaccharides represent a family of biopolymers whose diversity of structures offers a large spectrum of functional properties. Besides the use of these biopolymers in the exploitation of oil, in the food industry as texture agents, in agronomy, in the paper industry and many other domains, which have been known for years, scientists and investors are increasingly looking at the biological activities of these molecules and their applications in therapeutics. On this point, the scientific literature is regularly fed with data relative to the biological activity of these polysaccharides and their potential applications in different healthcare areas (anticoagulants, anti-inflammatories, anticancer, antiviral, dermal and bone regenerations, etc.) [COL 04].

The best-known bacterial EPS is xanthan, produced industrially for many years by the bacterium *Xanthomonas campestris pv. campestris*. Other polysaccharides are currently being developed in various industrial sectors such as gellane biosynthesized by *Sphingomonas paucimobilis*, dextrane (*Leuconostoc mesenteroides*), succinoglycane (*Agrobacterium radiobacter*), curdlane (*Agrobacterium* spp, *Rhizobium* spp), wellane, rhamnane, alginates (*Azotobacter vinelandii, A. chroococcum*, *Pseudomonas* spp) and bacterial celluloses (*Acetobacter* spp), cyclodextrines (*Bacillus* spp), pullulanes (*Aureobasidium pullulans*), herapin (*Escherichia coli* serotype K4 and K5) and hyaluronic acid [SUT 96]. These last two polysaccharides are particularly important in healthcare, especially low-molecular weight heparin as an anticoagulant, or hyaluronic acid as a biomaterial (filling product or scarring agent). However, today few bacterial polysaccharides of marine origin have experienced industrial development.

Acting on the fermentation conditions or even on the genetic material of microorganisms is a research field that may be completed by the discovery of new microorganisms able to synthesize these polysaccharides. The first approach does not yet appear very secure as the biosynthesis of EPSs is a complex process involving large numbers of enzymes, sometimes very specific to the synthesis of these macromolecules themselves. As part of the highlighting of new producing microorganisms, the atypical environments thus appear as special fields of investigation for this search for new species and biomolecules with original properties.

While EPSs influence the survival of bacteria with regard to the physical and chemical properties of their original habitat, those of the microbial mats, commonly called "kopara" (see section 3.4.2), can therefore prove to be interesting in this search for new producing microorganisms. A first study carried out in this way on a few microbial mats confirmed the presence of these EPSs causing some form of cohesion between the microbial communities, opening the possibility of finding producing microorganisms.

The creation of a collection of microorganisms is the first step of this protocol, which must lead to the discovery and production of new EPSs. This collection is the starting point of all research on microorganisms and the synthesis, in laboratory conditions, of innovative molecules such as exopolymers. For example, a collection belonging to PBT (Pacific Biotech SAS), located in Tahiti, currently holds over 1,500 bacterial isolates, mostly isolated from samples taken from these microbial mats.

The identification of EPS-producing bacteria starts with an initial screening step, followed by fermentation cultures in liquid medium at different scales. A notable change in the viscosity of the culture medium usually reflects the production of EPSs (Figure 3.20).

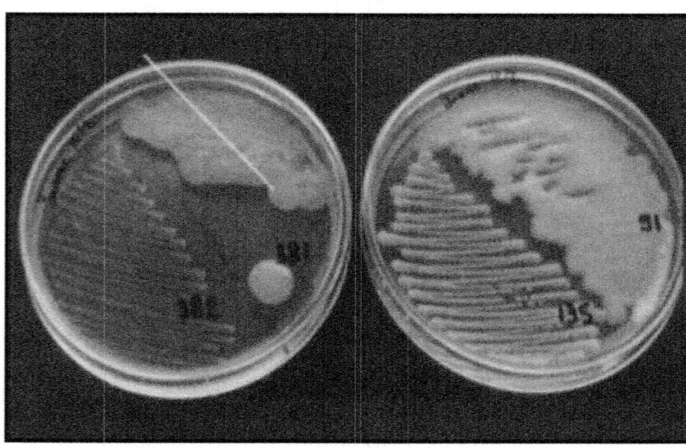

Figure 3.20. *Mucoid colonies characteristic of the production of exopolymers*

High-speed centrifugation results in the isolation of bacterial cells. Later treatment of the supernatant, either using cold alcohols or acetone, either with ultracentrifugation or successive dialysis, usually results in one or sometimes two exopolymers' isolation with very low rates of contaminants. The polysaccharide is then preserved from any hydration or other chemical or photochemical modification before undergoing analyses to determine the composition in monosaccharides and the presence of substituents (acetate, succinate, lactate, hydroxybutyrate, etc.). Indeed, the nature and position of polysaccharide chains is essential for some functions of these macromolecules. Analyses of these very complex macromolecules require, as for the previous molecules, a set of analytical chemistry methods, from simple gas chromatography to different techniques of mass spectrometry and multidimensional nuclear magnetic resonance.

EPS	Strains	Proteins	Neutral sugars	Uronic acids	Hexosamins	Sulfates	Substitutions
RA 1	Nd	3	24	15	16	–	Lac
RA 11	Nd	6	28	8	8	10	Ac, Pyr
RA 19	*Paracoccus* sp.	3	48	8	–	29	Ac
RA 29	*Vibrio* sp.	8	44	8	–	21	Ac
TE 7	*Pseudomonas* sp.	8	38	14	2	8	Lac
Tik 574	*Pseudoalteromonas* sp.	9	28	22		5	
Tik 650	*Alteromonas* sp.	4	47	10		12	
Tik 668	*Pseudoalteromonas* sp.	7	49	28		5	
Tik 725	*Alteromonas* sp.	7	20	17		–	
Mi 550	*Photobacter* sp.	5	34	18		–	
Mo 245	*Vibrio* sp.	2	11	27	30	–	Ac
Mo 203	*Alteromonas* sp.	4	46	20	–	–	

Table 3.1. *Examples of chemical composition of exopolysaccharides (EPSs) synthesized by bacteria isolated from microbial mats (kopara)*

While a few cyanobacteria isolated from these microbial mats have shown some ability to produce these exopolymers in laboratory conditions, the screening carried out on the collection of bacteria has highlighted a significant number of strains that could be used for this production. These bacterial EPSs are characterized by a large diversity in their chemical composition, with variable contents in neutral sugars (glucose, galactose, mannose, rhamnose and arabinose), acid sugars (galacturonic acid and glucuronic acid), acetylated or non-acetylated hexosamines (glucosamine and galactosamine) and substituting elements (acetate, lactate, pyruvate and sulfates) (Table 3.1).

Most of the polymers identified present an anionic character due to the presence of uronic acids (galacturonic and glucuronic acids), as well as the presence of substituting compounds like acetate or lactate groups. This anionic character provides the polymers with the potential for chelation of metallic ions, and important values have been observed for three of them, which were divalent ions like copper or nickel, or monovalent ions like silver. The applications are then at an environmental level as detoxification and decontamination agents of contaminated ecosystems, and also at the level of cosmetics and food, or even medical through the creation of organometallic complexes with antimicrobial properties.

Two polysaccharides, among those identified up to today, present high sulfate rates, with contents greater than 20%. One bacterium characterized as a *Pyrococcus sp.* (Strain RA19) produces a polysaccharide that is highly sulfated [RAG 04]. The presence of such a high level of sulfate groups is relatively rare in polysaccharides synthesized by bacteria, and is more frequently found in the polysaccharides of algae. The presence of these sulfates suggests interesting biological properties, in particular for applications in the cardiovascular domains as anticoagulants or antithrombotics [COL 01].

A second polysaccharide was identified as belonging to the genus *Vibrio* and is characterized by a high content of aminosugars, equivalent to those observed for sugar acids. This chemical composition goes back to other already known structures, used in a number of industrial sectors, such as hyaluronic acid. The chemical composition of this polysaccharide is comparable to that of a bacterium also belonging to *Vibrio* (*Vibrio diabolicus*), isolated from a deep hydrothermal ecosystem. This biopolymer has demonstrated, during various studies, properties of bone regeneration, as well as of skin scarring [ZAN 03].

To these specific properties can also be added those linked to the physico-chemistry of EPSs and their rheology[13] with applications in cosmetics as moisturizers or skin protectors.

3.4.2.2. *The polyhydroxyalkanoates*

Close to 300 million tons of plastic matter derived from the petrochemical industry are produced each year, 80% of which is aimed at everyday needs. Close to 17 billion plastic bags are used each year in France and more than 6 billion in Australia, and these are not the only users in the world (see [MON 14, Chapter 5]).

Biodegradable polymers can be used as alternatives to this type of synthetic material and the problems associated with their future in nature. The significance of biodegradable polymers in the world in 2000 was of 44,000 tons per year, only 0.12% of the plastics market. According to growth prediction for these polymers, including not only PHAs, but also polysaccharides and polymers of lactic acid (PLA), the market for this type of biopolymer could reach, in the next decades, between 10 and 20% of the plastic market. Moreover, the evolution of legislation on sustainable development, reducing greenhouse gases and CO_2 emissions, better management of waste, as well as a general desire to reduce dependence on imported primary matter, would plead in favor of the development of these biopolymers.

However, only bacterial PHAs can be considered to be 100% biodegradable and from renewable sources. These biopolymers accumulate within a bacterial cell in the form of granules, when a nutrient (source of nitrogen, phosphate, dietary elements, etc.) necessary for bacterial growth becomes a limiting factor in the presence of an excess of carbonated substrate (Figure 3.23). This

13 Rheology: study of the behaviors of polymers in solution.

synthesis is present in many bacteria (Gram + and Gram –), in some cyanobacteria, and very few eukaryotic cells. More than 300 microorganisms have been identified as being able to synthesize these polymers, but only some of them today have the potential to be developed industrially. The marine environment is part of the least studied ecosystems in terms of research on producer microorganisms. However, bacteria belonging to the genera *Halomonas*, *Alteromonas*, *Pseudoalteromonas*, *Pyrococcus* and *Vibrio* have recently been isolated from the marine environment, or ecosystems that are physically and chemically comparable to oceanic media [SIM 08a].

"Biodegradability" is a key word that characterizes these biopolymers. The PHAs are degraded in the environment with kinetics influenced by the physico-chemical conditions and microbiology of this environment. A lot of microorganisms possess depolymerase enzymes able to hydrolyze ester functions of the PHA into oligomers (OHA) and monomers soluble in an aqueous medium. These microorganisms, and some others, then metabolize these products of degradation into water and carbon dioxide.

The PHA can be divided into three categories:

– the *scl* (*short-chain length*) PHA made up of hydroxyalkanoid acids, with up to five carbon atoms;

– the *mcl* (*medium-chain length*) PHA, whose monomeric units contain 6–12 carbon atoms;

– the *lcl* (*long-chain length*) PHA, whose constitutive units contain 12–16 carbons (Figure 3.21).

The first group (*scl*) is rigid and brittle, and can be placed in the category of thermoplastics, which limits their applications, whereas the second and third groups (*mcl* and *lcl*) are part of the category of elastomers and adhesives and provide more possibilities for development (Figure 3.23).

Depending on the authors, there is a fourth category of PHA containing halogenated, epoxide, aromatic or other functions, classified under the term *fcl* PHA or functionalized PHA.

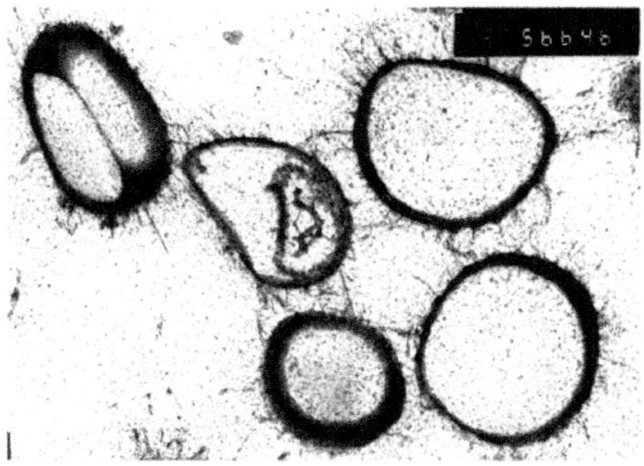

Figure 3.21. *Presence of granules of PHAs (polyhydroxyalkanoates) in the bacterial cells*

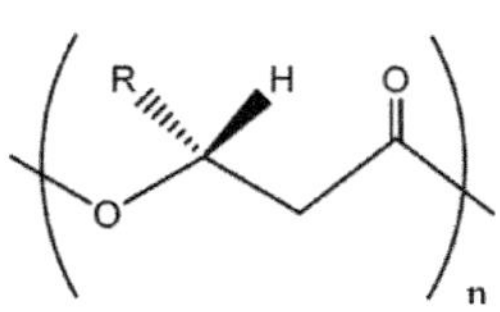

Figure 3.22. *General formula of a poly (β-hydroxyalkanoate) (PHA), where R = saturated/unsaturated/halogenated hydrocarbonated chain*

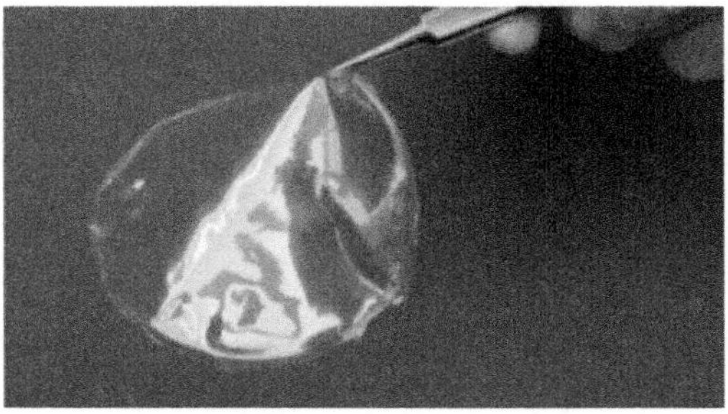

Figure 3.23. *Example of mcl PHA (in the form of a film) obtained by bacterial fermentation*

Biopolymers can be considered to be in their native state, or after chemical modification (epoxidation, addition of hydrophilic or hydrophobic groups, reticulation, functionalization, carboxylation, etc.). These possibilities of modification significantly increase the fields of application of these substances.

Due to their properly biodegradable character, PHAs were quickly presented as an alternative to what is commonly called "plastic", with evident applications in the domain of packaging. The possibility of producing copolymers to suit special needs allows reaching mechanical characteristics compatible with the notion of "packaging", but the production costs of these biopolymers, for a single application, make them unattractive for large-scale development at the moment.

Nevertheless, we can think that developments will occur in this general direction within the next few years. This may not be the case for the possible use of these biopolymers as biofilms in the food industry, considering the vast progress made both in production and methods of extraction of *mcl* PHA, as well as in the modification of these polymers (adding of mineral charges, "blends" or association of several PHAs or synthetic polymers). These modifications provide the finished products with the desired criteria, such as the fusion temperature, resistance to stretching, water and oxygen permeability.

Besides applications at the level of packaging, there are other opportunities for this type of biopolymer that could also experience development in the coming years, with regard to the evolution of legislation on environmental impact, as well as increased awareness in society and new protocols (REACH)[14]. Apart from the non-polluting advantages of these polymers, they also deal with a reduction of petrol stocks, and therefore of the products derived from petro-chemistry.

Long-chain PHAs could also be used as new "organic" adhesives, or even in body hygiene products. The use in aquaculture,

14 REACH: Regulation on the Registration, Evaluation, Authorization and Restriction of Chemicals.

horticulture, agriculture and even in cosmetics related to the control of the kinetics of the degradation of these biopolymers is envisaged for better phytosanitary efficiency and reduced environmental impact via a controlled release of the active compounds (herbicides, fungicides and antimicrobial molecules) or even in the preservation of sowed seeds. In this same direction, it is also possible to cite their potential in the domain of coatings and "antifouling" paints also using the controlled release of active products in the environment.

Finally, for the bioremediation and protection of our environment, it is useful to cite the potential of microorganisms to use phosphates as dietary elements in the synthesis of these PHAs, or even the coproducts of chemical and cosmetic industries (glycerol, PEG, volatile fatty acids coming from waste water treatment centers, etc.). An example can be given of polystyrene, for which waste represents more than 15 million tons for the USA. Transformed into styrene by pyrolysis, this later compound is then used by a bacterium as a source of carbon to produce *mcl* PHA.

Other properties can be exploited, such as the conception of water-resistant surfaces, or the possibility of creating stretchable films. Other applications are also possible in the conception of geotextiles or protective films to be used in the paramedical domain. However, the main applications today seem to be in the biomedical domains due to the use of specific properties of these polyesters: large diversity of structures and of physico-chemical and rheological characteristics, but also due to a better control of the production of these biopolymers and their degree of purity, as well as the many opportunities for modifications. The vectorization of the active ingredients orally or *via* patches and films, odontology, ophthalmology and dermal and bone reconstructions are major research fields throughout the world.

Like for EPSs, the issue remains of the point of searching for new strains that are more effective or that produce new polymeric structures isolated from atypical environments. Is there a link between the microbial mats of the *kopara* marshes and the presence of PHA-producing bacteria? Is the production of PHA, in nature, a response of microorganisms to conditions of nutritional and energetic disequilibrium, conditions that are reflected in the variability of the chemical and physical parameters of the ecosystems that make up these microbial mats? A significant number of microorganisms (> 50 isolates) have been isolated from different microbial mats with this ability to synthesize PHA and *de facto* have been the object of studies looking then at the nature of the biopolymers produced in laboratory conditions [SIM 08b, SIM 09].

mcl PHA, presenting interesting physical and chemical characteristics (low fusion temperature, resistance to stretching, etc.), have been produced at the lab scale by these bacteria from carbonated substrates as diverse as the coproducts of the petroleum (glycerol), cosmetic (polyethylene glycol-PEG) or even forest (levulinic acid). Some of these biopolymers are currently being evaluated for applications as much in the domain of packaging as in cosmetics and for the vectorization of active ingredients.

3.5. Bibliography

[ANO 11] ANONYME, *Lignes directrices pour l'accès aux ressources génétiques et leur transfert*, Editions Cirad, Inra, IRD, 2011.

[BIO 96] BIOGORNI L., PIETRA F., "Marine natural products applications", *Chemistry & Industry*, pp. 54–58, 15 January 1996.

[BLU 62] BLUMER M., "The organic chemistry of a fossil-I, the structure of the fringelite-pigments", *Geochimica & Cosmochimica*, vol. 26, pp. 225–227, 1962.

[COL 92] COLSON G., RABAULT B., LAVELLE F., *et al.*, "Mode of action of the antitumor compound girodazole (RP-49532A, NSC-627434)", *Biochemical Pharmacology*, vol. 43, pp. 1717–1723, 1992.

[COL 01] COLLIEC JOUAULT S., CHEVOLOT L., HELLEY D., *et al.*, "Characterization, chemical modifications and *in vitro* anticoagulant properties of an exopolysaccharide produced by *Alteromonas infernus*", *Biochimica & Biophysica Acta*, vol. 1528, pp. 141–151, 2001.

[COL 04] COLLIEC-JOUAULT S., HELLEY P.D., RATISKOL J., *et al.*, "Exopolysaccharides produced by bacteria isolated from deep-sea hydrothermal vents: new agents with therapeutic potential", *Pathologie Biologie*, vol. 52, pp. 127–130, 2004.

[DAM 89] D'AMBROSIO M., GUERRIERO A., DEBITUS C., *et al.*, "161. Corallistin A, a second example of a free porphyrin from a living organism. Isolation from the demosponge *Corallistes* sp. of the Coral See and inhibition of abnormal cells", *Helvetica Chimica Acta*, vol. 72, pp. 1451–1454, 1989.

[DAM 93] D'AMBROSIO M., GUERRIERO A., DEBITUS C., *et al.*, "On the novel porphyrins Corallistins B, C and E: isolation from the demosponge *Corallistes* sp. of the Coral sea and reactivity of their nickel derivatives (II) complexes toward formulating reagents", *Helvetica Chimica Acta*, vol. 76, pp. 1489–1496, 1993.

[DEF 94] DEFARGE C., TRICHET J., MAURIN A., *et al.*, "Kopara in Polynesian atolls: early stages of formation of calcareous stromatolites", *Sedimentary Geology*, vol. 89, pp. 9–23, 1994.

[DER 91] DE RICCARDIS F., IORIZZI M., MINALE L., *et al.*, "The Gymnochromes: novel marine brominated phenanthroperylenequinone pigments from the stalked crinoid *Gymnocrinus richeri*", *Journal of Organic Chemistry*, vol. 56, pp. 6781–6787, 1991.

[GUE 02] GUEZENNEC J., "Deep-sea hydrothermal vents: a new source of innovative bacterial exopolysaccharides of biotechnological interest?", *Journal of Industrial Microbiology & Biotechnology*, vol. 29, pp. 204–208, 2002.

[GUE 11] GUEZENNEC J., MOPPERT M., RAGUÉNÈS G., *et al.*, "Microbial mats in French Polynesia and their biotechnological applications", *Process Biochemistry*, vol. 46, pp. 16–22, 2011.

[HAL 13] HALL K.A., SUTCLIFF P.R., HOOPER J.N.A., *et al.*, "Affinities of sponges (Porifera) of the Marquesas and Society Islands, French Polynesia", *Pacific Science*, vol. 67, pp. 493–511, 2013.

[ISH 01] ISHIBASHI M., IWASAKI T., IMAI S., *et al.*, "Laboratory culture of the myxomycetes: formation of fruiting bodies of *Didymium bahiense* and its plasmodial production of makaluvamine A", *Journal of Natural Products*, vol. 64, pp. 108–110, 2001.

[KEI 11] KEIVAN Z., BOON-TEONG T., SING-SIN S., *et al.*, "*In vitro* antiviral activity of fisetin, rutin and naringenin against dengue virus type-2", *Journal of Medicinal Plants Research*, vol. 5, pp. 5534–5539, 2011.

[KOR 05] KORNPROBST J.M., *Substances naturelles d'origine marine: chimiodiversité, pharmacodiversité, biotechnologies*, Lavoisier, Paris, 2005.

[LAI 98] LAILLE M., GERALD F., DEBITUS C., "*In vitro* antiviral activity on dengue virus of marine natural products", *Cellular and Molecular Life Science*, vol. 54, pp. 167–170, 1998.

[LAU 05] LAURENT D., BAUMANN F., BENOIT A.G., *et al.*, "Structure activity relationship of dengue antiviral polycyclic quinines", *East Asia Journal of Tropical Medicine and Public Health*, vol. 36, pp. 3516–3521, 2005.

[MAN 95] MANCINI I., GUELLA G., DEBITUS C., *et al.*, "Novel naamidine-type alkaloids and mixed-ligand zinc (II) complexes from a calcareous sponge, *Leucetta* sp., of the Coral Sea", *Helvetica Chimica Acta*, vol. 78, pp. 1178–1184, 1995.

[MAN 05] MANCUSO NICHOLS C.A., GUEZENNEC J., BOWMAN J.P., "Bacterial exopolysaccharides from extreme marine environments with special consideration of the southern ocean, sea ice, and deep-sea hydrothermal vents: a review", *Marine Biotechnology*, vol. 7, pp. 253–271, 2005.

[MAN 06] MANCINI I., GUELLA G., FROSTIN M., *et al.*, "On the first polyarsenic organic compound from nature: arsenicin A from the New Caledonian marine sponge *Echinochalina bargibanti*", *Chemistry, A European Journal*, vol. 12, pp. 8989–8994, 2006.

[MON 14] MONACO A., PROUZET P. (eds.), *Vulnerability of Coastal Ecosystems and Adaptation*, ISTE, London, and John Wiley & Sons, New York, 2014.

[PAT 05] PATERSON I., ANDERSON E.A., "The renaissance of natural products as drugs candidate", *Science*, vol. 310, pp. 451–452, 2005.

[RAG 04] RAGUENES G., MOPPERT X., ROCHERT L., *et al.*, "A novel exopolymer-producing bacterium, *Paracoccus zeaxanthinifaciens* subsp. *payriae,* isolated from a "Kopara" mat located in Rangiroa, an atoll of French Polynesia", *Current Microbiology*, vol. 49, pp. 145–151, 2004.

[RIN 88] RINEHART K.L., KISHORE V., BIBLE K.C., *et al.*, "Didemnins and tunichlorin: novel natural products from the marine tunicate *Trididemnum solidum*", *Journal of Natural Products*, vol. 51, pp. 1–21, p. 624, 1988.

[SAL 07] SALEEM M., ALI M.S., HUSSAIN S., *et al.*, "Marine natural products of fungal origin", *Natural Products Reports*, vol. 24, pp. 1142–1152, 2007.

[SIM 08a] SIMON-COLIN C., RAGUENES G., COZIEN J., *et al.*, "*Halomonas profundus* sp. nov., a new PHA-producing bacterium isolated from a deep-sea hydrothermal vent shrimp", *Journal of Applied Microbiology*, vol. 104, pp. 1425–1432, 2008.

[SIM 08b] SIMON-COLIN C., RAGUENES G., CRASSOUS P., *et al.*, "A novel mcl-PHA produced on coprah oil by *Pseudomonas guezennei* biovar. *tikehau*, isolated from a "kopara" mat of French Polynesia", *International Journal of Biological Macromolecules*, vol. 43, pp. 176–181, 2008.

[SIM 09] SIMON-COLIN C., ALAIN K., RAGUÉNÈS G., *et al.*, "Biosynthesis of medium chain length poly(3-hydroxyalkanoates) (mcl PHAs) from cosmetic co-products by *Pseudomonas raguenesii* sp. nov., isolated from Tetiaroa, French Polynesia", *Bioresource Technology*, vol. 100, pp. 6033–6039, 2009.

[SOL 04] SOLOMON C.E., MARGARVEY N.A., SHERMAN D.H., "Merging the potential of microbial genetics with biological and chemical diversity. An even brighter future for marine natural products drug discovery", *Natural Products Reports*, vol. 21, pp. 105–121, 2004.

[SUT 96] SUTHERLAND I.W., "Biotechnology second", in REHM H.J., REED G. (eds.), *Products of Primary Metabolism*, VCH, Weinheim, vol. 6, p. 613–657, 1996.

[TIC 12] TICHÝ M., POHL R., XU H.Y., *et al.*, "Synthesis and antiviral activity of 4, 6-disubstituted pyrimido[4,5-b]indole ribonucleosides", *Bioorganic and Medicinal Chemistry*, vol. 20, pp. 6123–6133, 2012.

[VER 09] VERON J.E.N., DEVANTIER L.M., TURAK E., *et al.*, "Delineating the coral triangle *Galaxea*", *Journal of Coral Reef Studies*, vol. 11, pp. 91–100, 2009.

[WAN 10] WANGUN H.V.K., WOOD A., FIORILLA C., *et al.*, "Gymnochromes E and F, cytotoxic phenanthroperylenequinones from a deep-water crinoid, *Holopus rangii*", *Journal of Natural Products*, vol. 73, pp. 712–715, 2010.

[WOL 08] WOLKENSTEIN K., GLUCHOWSKI E., GROSS J.H., *et al.*, "Hypericinoid pigments in millericrinids from the lower Kimmeridgian of the Holy Cross mountains (Poland)", *Palaios*, vol. 23, pp. 773–777, 2008.

[ZAN 03] ZANCHETTA P., LAGARDE N., GUEZENNEC J., "Systemic effects on bone healing of a new hyaluronic acid-like bacterial exopolysaccharide", *Calcified Tissue International*, vol. 73, pp. 232–236, 2003.

4

Marine Renewable Energies

4.1. Introduction

To fight against the effects of climate change, and with the goal of engaging or supporting a drastic reduction in the release of greenhouse gases on a national, European or international level, reduced use of fossil fuels is necessary. Renewable energy (RE) sources, such as biomass, hydraulics, thermal or photovoltaic (PV) solar energy, geothermal, terrestrial wind power and marine energies, represent a huge and limitless potential, available in all parts of the world. Although non-negligible, they cannot currently replace the energy provided by fossil fuels. However, there is currently a general consensus on the necessity to develop these sources. The area of marine renewable energies (MREs) is currently emerging. Technological and non-technological issues need to be resolved to allow the various industrial specialties to reach a level of technical and economic profitability, which, depending on the level of development of the area, could be achieved between 2020 and 2030.

Chapter written by Michel PAILLARD, Bernard MULTON and Marc BŒUF.

4.2. The energetic context and the stakes in the development of renewable energies

4.2.1. *Energy resources, reserves and final energy, definitions and concepts*

First of all, it is essential to distinguish nonrenewable energy sources, which started to be exploited on a massive scale in the 19th Century, from those that are renewable. Next, the difference must be made between primary energy sources, which are those that can be extracted from our natural environment, from exploitable energy forms, called final, to satisfy our needs, and finally energy services [MUL 11]. Figure 4.1 proposes a synthetic vision of these three energy levels, from primary resources to energy services.

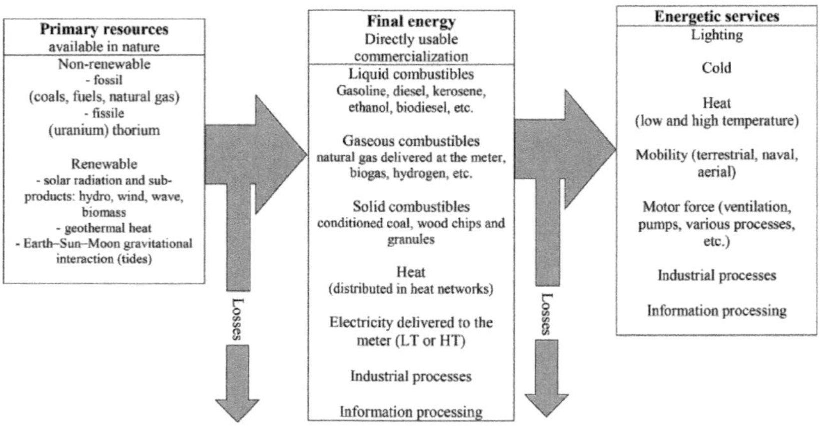

Figure 4.1. *Primary resources to final energy to satisfy services*

Figure 4.2 shows the global distribution for 2011 of the primary resources consumed (more than 13 Gtep[1]) and the final energy commercialized (8.9 Gtep). The losses are due to the extraction of

1 Gtep = gigatep = 1 billion of petrol equivalent, or an energy value of 11,600 TWh (1 TWh = 1 terawatt hour = 1 billion kWh).

energetic primary matter, its transformation into exploitable energy and its distribution. Currently, it is electricity, mainly produced from nonrenewable resources with mediocre efficiency, which is the source of most of the losses in transformation. It should be noted that currently, over 86% of our primary resources are derived from non-renewable sources.

Renewable energy, [CHA 97] on the other hand, is derived from continuous and limitless (on our timescale) sources. These include solar radiation and its derivatives (hydraulics, wind, swells, thermohaline circulating ocean currents), geothermal heat and the gravitational effects of the Earth–Moon–Sun interaction (which notably leads to tidal effects). On the scale of the entirety of the terrestrial and oceanic surfaces, these renewable resources represent a total potential reserve of a massive 100,000 Gtep, roughly, or around 8,000 times the annual global consumption of primary energy for the whole of humanity, more than two thirds of which are located in oceanic zones, due to the huge areas that they occupy. An annual quantitative summary of the different forms of RE is described in Figure 4.3.

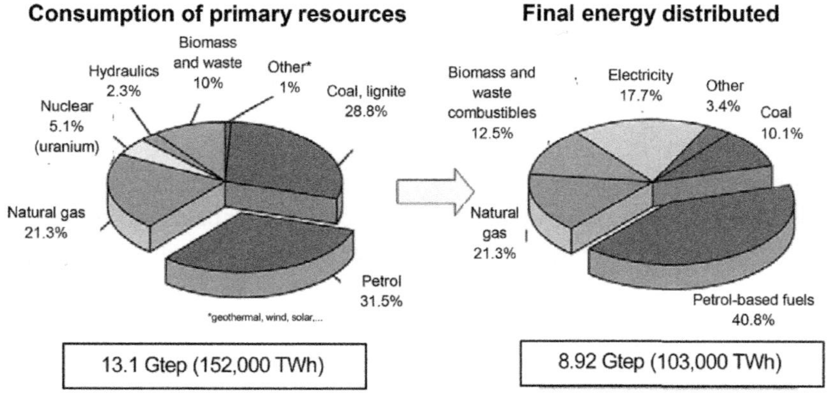

Figure 4.2. *Global consumption in 2011 of the primary resources of the final energy distributed [IEA 13]*

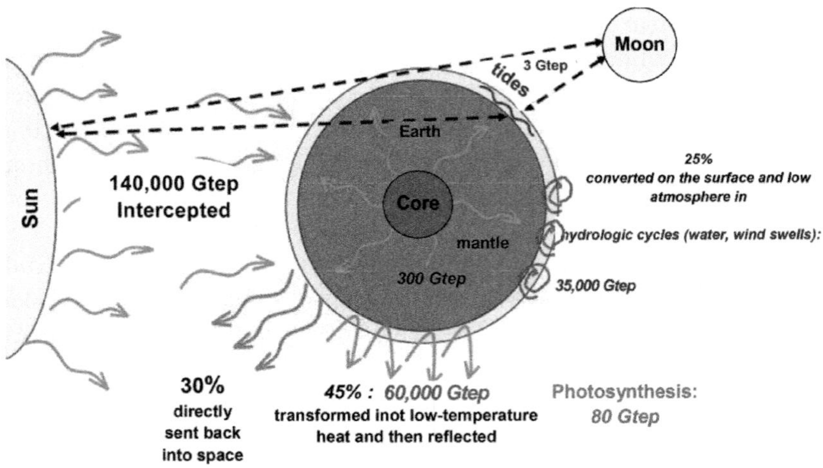

Figure 4.3. *Summary of the annual planetary fluctuations of renewable energy that make up all renewable resources*

For comparison, all the primary nonrenewable resources stored in the Earth's crust represent a reserve of between 2,000 and 5,000 Gtep for slightly less than 1,000 Gtep of proven primary reserves (of which less than 60 correspond to uranium, considering the typical fission of isotope 235). Let us note that the reserves constitute the resources that have been discovered (proven, identified, or even "reasonably certain of", to borrow the terminology used in the case of uranium). The proportion can increase in function of new discoveries and improvements in recuperation technology, but it can in no case be larger than the resources. The reserves are commonly (and sometimes mistakenly) quantified in number of years by expressing the ratio, called R/P, of the identified reserves over annual production. Thus, for petroleum, the R/P ratio has remained at more or less 40 years for several decades, but it is clear that, despite access to nonconventional resources (oil sands, shale gas, etc.), this cannot last. Indeed, the reserves cannot be greater than the resources and the R/P ratio is in the order of a hundred years, all forms of fossil fuel combined; the main problem is first of all linked to the amount of carbon released into the troposphere.

First let us return to the forms of final energy, those that are directly usable for different needs, whether residential, industrial, mobility, etc. As a first example, let us consider the case of commercial combustible liquids, which are majorly used in the transport sector. They are currently produced from fossil combustibles with a relatively low amount coming from cultivated biomass (biofuels). This energy comes from solar radiation but also from fossil fuels used for mechanical energy and for various inputs.

Let us now consider electricity, as a second example. In 2012 (Figure 4.4), 79.2% was produced by nonrenewable energy resources (fossil and nuclear), first converted into heat, and then into electricity with low levels of efficiency (30–60%) via thermodynamic cycles. The 20.8% comes from 78% of hydraulics and about 11% from wind power. However, with very high growth rates, wind power production (average of 26% per year from 2002 to 2012) and solar production (average of 50%) should become the majority well before 2030. Let us note that percentage occupied by nuclear production has been dropping constantly for several years. With a proportion of 10.9% (Figure 4.4) of electricity produced and a contribution of electricity at 17.7% of total final energy (Figure 4.2), it turns out that the contribution of uranium to the provision of world in final energy is 2%, which puts its importance into perspective.

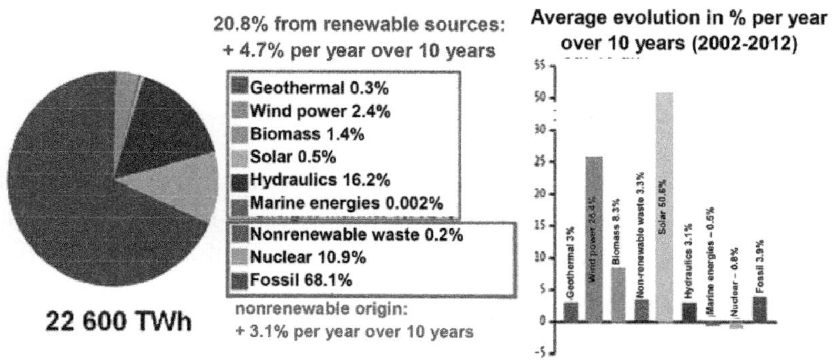

Figure 4.4. *Global energy production in 2012 and average rates, over ten years, of growth for each source [OBS 13] (see color section)*

To compare energy production systems, in this case of electricity, we have become used to considering energy yields without considering all of the lifecycle and the renewable or nonrenewable nature of the resources consumed. Thus, a PV single crystal silicon generator has a nominal efficiency of around 20%, lower than that of nuclear power plant (33%), which converts uranium into heat (fission) and then into electricity via a vapor turbine, which can appear an advantage for nuclear. But if we look at the whole lifecycle of these two devices and we calculate the amount of nonrenewable primary energy consumed (corresponding to resources that are definitely destroyed) to produce 1 kWh, it would appear that PV generation is more effective, and this is all the more true when the system is located in good sunshine conditions. For example, we currently consume around 5 kWh to create a PV production capacity of 1 W using multicrystal silicon technology (complete system) under a sunshine exposure of 1,000 h per year (Northern Europe), producing over 25 years around 25 kWh (25,000 × 1 W), or five times more than the energy used to create it. This is even better with thin layer technologies and with wind turbines. Thus, the notion of lifecycle yields, defined on the basis of the consumption of nonrenewable primary energy, allows us to compare solutions in terms of sustainable development. Figure 4.5 allows us to estimate a global energy notion of the lifecycle of an energy converter, and shows the different consumption posts, as well as the renewable or nonrenewable origins of the primary energy. The gray energy corresponds to the energy required to create, recycle and maintain the system in working condition over the whole of its lifespan.

We can also consider the greenhouse gas emission over the whole lifecycle and bring back to the kWh. In this case, nuclear production would be more effective than PV productions if the energy used to produce the PV system is carbonated. But if we look at radioactive emissions over all of the lifecycle, this would obviously be the other way around. Today, software is available to conduct lifecycle analyses (LCA) and carry out multi-criteria comparisons (on the basis of different environmental impacts). The processes of conversion of renewable resources into electricity then appear as having less impact [SAT 11].

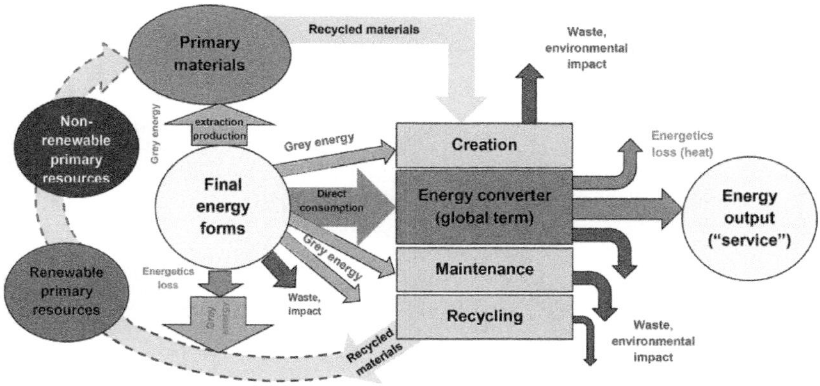

Figure 4.5. *Primary energy (renewable and nonrenewable) required to satisfy an energy service over all of the lifecycle of an energy converter*

The new technologies for the conversion of MREs must be examined using LCA to effectively evaluate their potential for contributing to sustainable development.

4.2.2. *Global, European and French contexts: the stakes*

The appearance of the concept of sustainable development at the end of the 20th Century has led humanity to take into consideration, on the scale of the planet, the major challenges faced by its extraordinary expansion. It is by conducting massive extractions in nonrenewable fossil fuels, and to a certain extent in fissile materials, that an unprecedented acceleration of population growth has been possible. Until the end of the 18th Century, the metabolism of human societies was mainly founded on RE sources: wood, hydraulics and wind. The access to nonrenewable stores of coal, petrol and gas caused an unprecedented frenzy of activity that resulted in a snowballing of the consumption of energy – mainly nonrenewable (Figure 4.6).

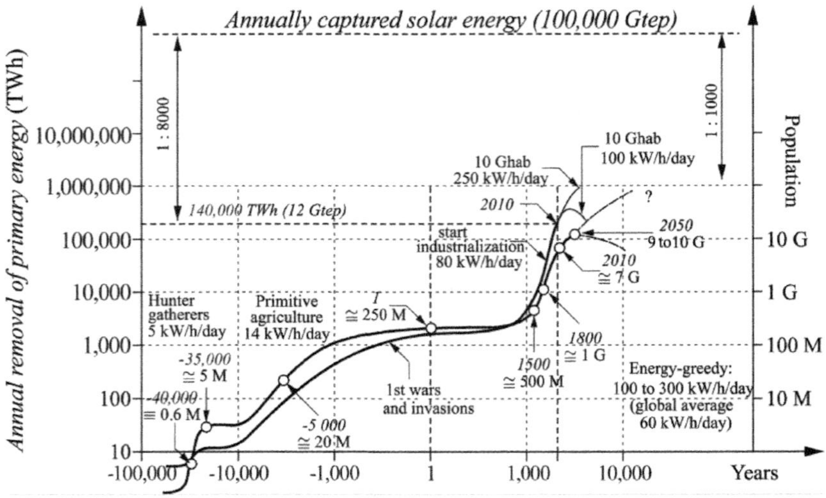

Figure 4.6. *Growth of human population and associated consumption of primary energy since prehistoric times*

One of the major challenges that faces humanity is that of global warming. Around 60% of greenhouse gas (GHG) emissions of human origin are due to the consumption of fossil fuels, and therefore energy-related activities. To this can be added the rarefaction that leads to an increase in environmental damage during their extraction as well as violent effects to the global economy caused by market tensions. Uranium, although not a big contributor to GHG emissions, does not escape rarefaction of resources as well as economic and geopolitical tensions. It is also the cause of other pollutions that are just as worrying as GHG emissions in terms of the irreversibility of their consequences.

Over the last million years, the concentration of CO_2 (main GHG) in the atmosphere has oscillated between 180 and 280 ppm[2], the periods of low concentration corresponding to ice ages. Since the 20th Century, we have left this range of variation. In 2012 we crossed the 390 ppm line, which represents a violent impact on our climate

2 ppm = parts per million.

regulation system. In its latest report (2007) [IPC 07], the IPPC[3] advised a limit of 450 ppm of CO_2 concentration to contain global warming inside a bracket of a few degrees, supposedly still reversible. For this, a drastic decrease in emissions is necessary, of around 13 Gteq CO_2[4] per year by 2050, while current emissions are greater than 50 Gt, for seven billion inhabitants; we can expect to be around 10 billion in 2050, a large part of which will have access to increasingly energy consuming lifestyles. Assuming this size of global population, an average of 1.3 tons CO_2 equivalent per person would have to be achieved, which would require an impressive amount of effort, in particular for rich countries. Figure 4.7 shows the values of the emissions per person, which, in 2004, went from 2 to 27 tons CO_2 equivalent depending on the region; these averages also mask strong regional disparities (sub-Saharan Africa with less than 100 kg and more than 50 tons for some petroleum-producing countries). The famous "factor 4", which is often heard in France, corresponds to the effort that needs to be made to reduce emissions in line with the global reduction that needs to be seen by 2050.

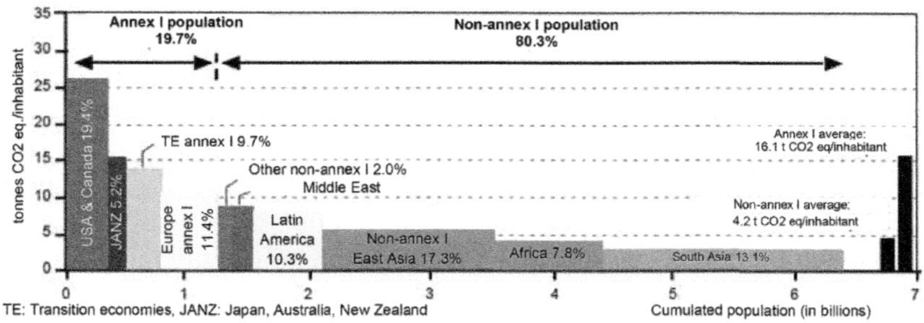

Figure 4.7. *Emissions (2004) of GHG (in eqCO_2) per inhabitant according to the different categories defined by the IPPC [SOS 13]*[5]

It is in 1992 that the United Nations Framework Convention on Climate Change was adopted, with the objective of pushing nations to

3 Intergovernmental Panel on Climate Change

4 tons CO_2: tons CO_2 equivalent, unit of power of global warming allow the combining of the contributions of all GHG in the equivalent of carbon dioxide.

5 The "annex I TE" are the countries of Eastern Europe, including Russia.

stabilize the concentration of GHG in the atmosphere. In 1997, a majority of States signed the Kyoto Protocol, in which they promised to decrease their GHG emissions, with a global objective of a reduction of 5% of emissions in 2012 compared to 1990s levels. This global objective was then transposed to national objectives for the nations ratifying the protocol, i.e. the vast majority with the exception of some like the USA (20% of all emissions in 2007) and Australia, who are among the biggest emitters. The protocol only really started being applied in 2005. The signatory countries are placed into two different categories (Figure 4.7): annex 1 (developed) and non-annex 1 (developing countries), which are not expected to reduce their emissions, considering their right to development (this includes China, with 22% of total emissions in 2007).

Political negotiations are difficult. In Copenhagen (2009), new commitments for 2020 ("post-Kyoto") have been made by only some of the initial countries, but without a global objective, and the latest annual conferences (Cancun, Durban and Doha in 2012) did not succeed in coming up with the absolute promises that must be met to stop the spirally of global warming. Currently, it is the European Union (UE-27) that has set itself the biggest limitations, with a reduction of 20% for 2020 (still in relation to 1990). France has agreed to reduce its emissions to a quarter by 2050 during the *Grenelle de l'environnement*.

To achieve these reductions, it is necessary to considerably reduce energy consumption through greater effectiveness (technological) and sobriety (organizational and behavioral) as well as through the use of renewable resources. On this basis, without using nuclear energy, the *negaWatt* scenario of 2011 proposes a reduction in emissions by a factor of 16 for 2050 [SAL 12].

It is in this context that renewable marine resources would contribute to a global future of mixed energy sources, with a high percentage of renewables to produce electricity (easily transportable and convertible) or even combustibles (hydrogen, synthetic methane, etc.).

4.3. The place of marine energies in renewable energies

4.3.1. *Marine renewable energies (MRE)*

4.3.1.1. *Different energy resources*

MREs are not only mechanical energies from the movement of bodies of water, but also physico-chemical energies from temperature differences or salinity differences, and finally also energies from biotechnology like the production of biofuels from microalgae. Thus they cover tidal energy, currents, waves, temperature gradients between the surface and the depths (Ocean Thermal Energy Conversion – OTEC), salinity gradients and marine biomass. Wind energy at sea is not specifically a marine energy source, but we will consider it as such due to similarities of the operation methods with those used in marine energies. Offshore wind technology does share environmental conditions with MRE as well as design, offshore installation, operation and maintenance conditions and dismantling. Finally the use of cold water from the ocean depths (Sea Water Air Conditioning – SWAC) allows considerable savings to be made in terms of electricity. Energy costs, the safety of getting supplies to insular areas, notably in French overseas territories, and the increasing growth of global coastal populations are all arguments in favor of the use of energy harnessed from the sea. The development of OTEC could lead to the replacement of fossils, like coal, in the energy mix of certain overseas territories.

4.3.1.2. *Natural potentials and exploitable potentials*

The issue of the potentials of different marine energy resources is recurrent, as many publications cite estimations without mentioning the methods used to get them. This leads to interpretations that can harm the credibility of marine energies. After all, a distinction must be made between the natural resource, commercially useless, and the technically exploitable potential.

This notion of *technically exploitable potential* appears therefore more relevant, as long as the temporal aspect of this estimation is highlighted. Indeed, it is clear that these technically exploitable potentials will change over time, in function of the evolution and the level of technologies, of the evolution of other uses of the marine space, of regulation, notably in terms of environment protection, and finally of societal acceptability. We will undoubtedly witness changes in the perception of the need to develop renewable energies, and the acknowledgement of their eventual limitations in an energetic and economical context, linked to the rarefaction of fossil resources, to climactic limitations, to many unresolved questions on electronuclear industry, and therefore the inevitable increase the costs of "conventional" sectors. This necessary energy transition is already the topic of many debates in 2013.

4.3.1.3. *Estimations of potentials by sectors*

A global summary of these resources (Table 4.1) is proposed in the foreword of the work Marine Renewable Energy Handbook published in 2012 [MUL 12], which highlights that "figures are only given as an approximation. The orders of magnitude of the exploitable and recoverable parts, as well as the yields considered to evaluate the exploitable electrical energy can be questioned. Their exact origin, as well as the associated explanations, can be found in [MUL 09]". To give a reference, the annual production of electricity was around 22,000 TWh.

Annual energies 1 000 TWh	Solar oceans	Continuous ocean currents	Marine biomass	Hydrological cycles			Tides	
				Offshore wind	*Wave*	*Osmotic*	*Dams*	*Currents*
Global estimate	400,000	?	450	>> 500	80	30	22	
Primary exploitable part	4,000 (1/100th ?)	1 ?	4 (1/100th?)	100	2	0.3 ?	0.6	1.8
Recoverable electrical energy	80 (η 2%)	0.3 ? (η 30%)	1 (η 25%)	37 (η 37%)	0.75 (η 35%)	0.1 ? (η 30%)	0.6	0.9

Table 4.1. *Order of magnitude of the resources of renewable energies and their exploitable part in primary energy, and electrical, with production methods for which the efficiencies (η) are given [MUL 12]*

Some recent publications also propose estimations by resource, sometimes based on precise methodology, like for example Chapter 6 of "Ocean Energy" and Chapter 7 of "Wind Energy" from the *Special Report of the Intergovernmental Panel on Climate Change* (IPCC) "Renewable Energy Sources and Climate Change Mitigation" [IPC 12]. Other recent publications from the *World Energy Council* [WEC 10] and the IEA-OES [IES 12] can also be downloaded and referred to by the reader. Other estimations are provided there, like 44,000 TWh/year for global OTEC resource or 29,500 TWh/year for the global wave energy resource. Finally, as an example, Figure 4.8 shows the geographical distribution of the temperature difference between waters pumped at 20 m and at 1,000 m. The OTEC resource concerns zones where the difference is greater or equal to 20°C.

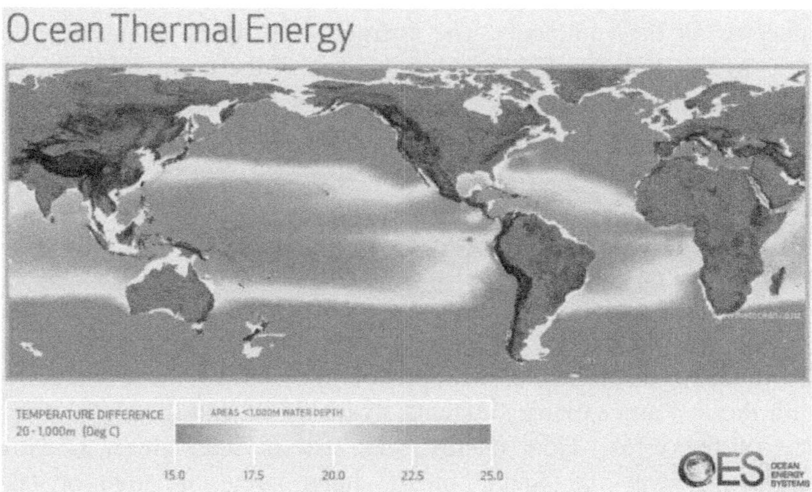

Figure 4.8. *Global OTEC resource [OES 12] (see color section)*

For France, there is no study that uses a reliable and consistent methodology. Estimations have been put forward for some MRE in recent years, like the exploitable tidal current potential, between 5 and 14 TWh/year, or like that of wave energy converters of 40 TWh/year, corresponding to 1/10th of the estimated natural potential. For offshore wind turbines, including floating ones, a first estimation is proposed of around 200 TWh/year. With regards to the exploitable potential of OTEC in French overseas areas, it is probably very high, but also undoubtedly limited (low levels of local need of the potential and environmental limitations in the case of large-scale exploitation. These elements confirm that the estimation of these potentials remains today a research subject in itself.

4.3.2. *The specificities of MRE and their place in the energetics mix*

4.3.2.1. *Global distribution, predictability*

MREs also distinguish themselves by their spatio-temporal specificities. While the OTEC can only be of interest in the inter-tropical zone, the geographical distribution of wind and wave energy concerns all the continents. The energies from tides are localized at areas of the globe where the configuration of the coast amplifies the size of the tidal wave. They are therefore by definition localized. The energy of salinity gradients can be exploited anywhere with two sources of freshwater and seawater. However, coastal competition considerably reduces these locations in territories that are strongly urbanized, as well as the knowledge of the environmental impacts of waste.

Two marine energy resources can be considered as capable of continuous production, or basic production, and are those associated with thermal and salinity gradients, available around the clock, every day of the week. Tidal energy, like that in tidal currents, while fluctuating over the course of the lunar cycle, is, give or take metocean conditions, globally predictable. Thermohaline circulation, which generates the oceanic currents, also presents, locally, this advantage, with well-known seasonal variations. Finally, the energy of wind and waves is highly variable by nature. But tools developed in recent decades, to refine their prediction in the short-term, means that they can be classed in the category of semi-basic energy sources, and this prediction allows power plants to engage in short-term production, vital for the grid management to balance production in function of demand.

4.3.2.2. *Conditions for a contribution to the energetics mix*

The first condition for the development of MRE is to utilize the strong institutional support, accompanied by industrial investments, from those in the sector. This has particularly been the situation for British companies for over 10 years. In France, this support has been evident since the *Grenelle environmental initiative* and the *Grenelle Oceans initiative* in 2008 and 2009. But the large-scale development of MRE, a requirement for a significant contribution to the energy mix[6], in addition to the resolution of several technological and non-technological (notably environmental) issues, entails the establishment of consultation and planning tools aimed at favoring the establishment of a real national and European strategy for the development of MRE in the medium- and long-term. It also requires access to socio-economically acceptable storage conditions compatible with an MRE economy. This is notably the case for OTEC, where the resource is located in zones of low-demand in terms of consumption. These conditions are not currently met, as is needed for large-scale development, but support for R&D, and the establishment of demonstrators and pilot farms as part of public–private partnerships, started in 2009, should result in the emergence of these marine energy sectors during this decade.

4.3.3. *The four stakes of the development of MRE in France*

To describe the economical challenges of MRE development, first we must remind ourselves of the different stakes involved:

– an ecological stake, for sustainable development by contributing to a reduction of greenhouse gases generated by sources of electricity derived from fossil fuels, and a reduction of other pollutants such as radioactive emissions. Obviously, this stake is associated with a necessary energy efficiency and a reduction of energetic needs through better management of consumption;

– an energetic stake, by contributing to the energy mix. This is an increasingly urgent necessity in the context of the imminent depletion of nonrenewable resources. If in 2020, the objective of the

6 The energy mix must permit a transition toward a sustainable energy system, able to supply the demand, while conserving the environment.

Grenelle de la mer, of 6,000 MW, is not achieved, the following decade must establish a significant development of MRE, notably in oversea territories;

– an industrial and economical stake, for the development of complex industrial sectors, generating local and specialized jobs. In France, the potential for the development of MRE is significant due to abundant exploitable natural resources, and French companies with world-renowned expertize in energy, naval construction and offshore activities. Finally, the level of excellence reached by French research in the sciences and marine engineering will contribute to the development effort. This monumental effort will need to be financed, and will represent a great potential for economic growth;

– a social stake, since the energy transition will not be possible without citizen involvement. The time when changes were imposed without discussion is long gone and social sciences must now be part of the debate. Consultation forums are set up. Awareness of the geopolitics of energy is much more present today. Alternative development scenarios are proposed and should contribute to the definition of public politics in the matter of energy, notably at the European level.

4.3.4. *Technological diversity*

The previous chapters have presented the richness in diversity of RE resources available at sea. The technological solutions for their conversion will in turn become more diversified and must also be adaptable, within the different resources, to the high variability of the natural environment and the geographical and socio-economical contexts of the sites being equipped.

These systems are at different stages of technological maturity as represented by the following graph (Figure 4.9).

The following sections present a summary of the main types of equipment used. For more details, readers can refer to [MUL 12].

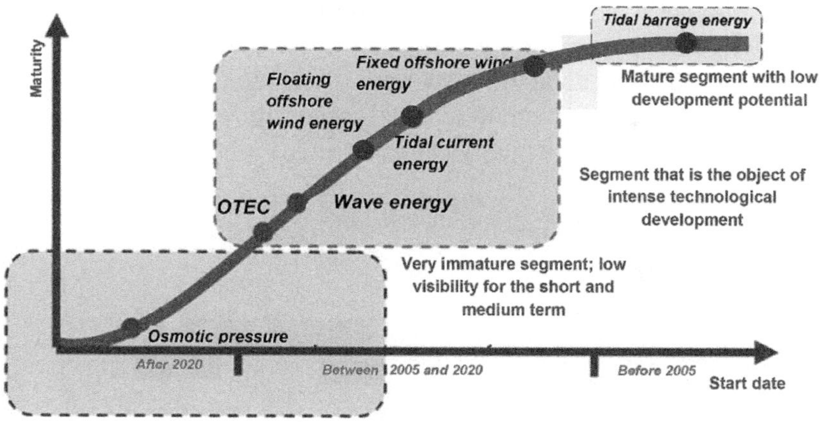

Figure 4.9. *The development of different technologies (France Energies Marines/Indicta, 2012)*

4.3.4.1. *Tidal barrage*

The recovery of the potential energy (increase in the water level) from tides on the coast has existed for a long time. Tidal mills are historical examples of this, found notably on the west coast of France.

At the beginning of the 1960s, a proactive industrial program resulted in the installation of the first high-power tidal barrage power plant (240 MW) at the mouth of La Rance in Brittany (Figure 4.10). This remarkable equipment, operated by EDF, has since been producing more than 500 GWh per year and provides around 2.5% of the electrical needs of this region. Today, a new tidal barrage power plant has been installed in South Korea (Sihwa site), in 2012, and its installed power is of the same order of magnitude (power 254 MW for an annual production of around 550 GWh).

The large-scale recovery of the potential energy of tides is the subject of new studies at the international level. Thus, South Korea, but also the UK, are preparing for this development, while taking into account the inherent difficulties linked to these large coastal installations, in particular in terms of the impacts on the environment and on coastal production practices. Many studies have been carried out for the projects found in the estuaries of rivers Severn and Mersey.

Tidal energy at coasts can therefore provide huge production over very long time periods, but is limited to a few coastal sites capable of satisfying the technical and economical parameters, but also the criteria of social and environmental acceptability, as well as the financial aspect of these highly capitalist endeavors.

Figure 4.10. *Tidal barrage power plant of La Rance (© EDF – DR)*

To overcome these coastal limitations, in the medium-term, offshore systems could be developed so as to reconstitute a "tidal power reservoir" close to the coasts (artificial lagoons), assuming that the increase in cost of transferring energy toward the land would not be excessive, the main impact being then carried over to other uses.

4.3.4.2. *Tidal current turbines*

Tidal current turbines capture part of the kinetic energy of tidal currents. These are systems of various technologies that have the advantage of benefiting from quite dense and predictable energy, but which is localized to only some sites.

A tidal current turbine can be likened to an underwater wind turbine whose rotor is turned by the current. Tidal current turbines are, for the most part, submerged, reducing impacts on use and on the landscape, but they must be designed to withstand long-term immersion, sometimes in hostile zones to limit the frequency of sub-marine maintenance, which is very costly. Several of these systems are in their demonstration stage at sea, in the pre-industrial phase. The main technological families of tidal current turbines can be classified depending on their type of foundation or their turbine type. All the illustrations are taken from the website (www.aquaret.com):

– foundation type: gravitational, mono or multiple piles, floating;

– turbine type: axial flow (horizontal axis, Figure 4.11, example SeaGen), transverse flow (vertical axis, Figure 4.12, example HydroQuest, project from a French start-up), other: impeller, oscillating profile (Figure 4.13), etc.

Figure 4.11. *Axial flow tidal current turbines–SeaGen turbine (© SIEMENS/MCT 2008)*

Figure 4.12. *Transverse flow tidal current turbines*

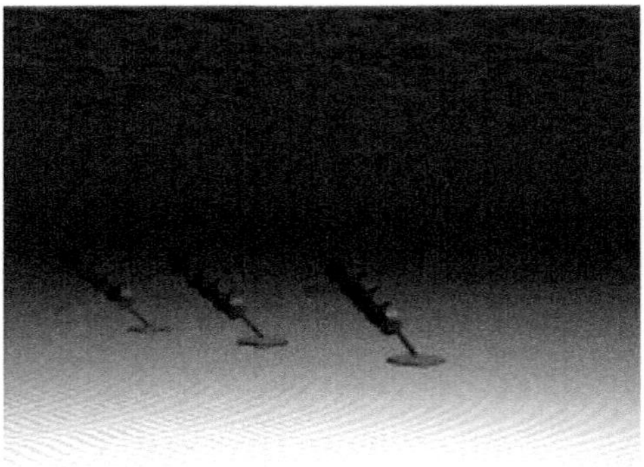

Figure 4.13. *Oscillating profile tidal current turbines*

Some systems are equipped with nozzles so as to increase the speed of the current at the level of the turbine (by Venturi effect) and to reduce its size (Figure 4.14), such as the MegaWattBlue in France. However, the concept suffers from the increase of stresses on the structures.

Today, axial flow tidal current turbines with a small number of blades are the most developed. Their unitary power goes from several hundred kW to more than a MW.

In the longer-term, systems could be developed to capture the energy of less dense currents, further from the coasts, like the great ocean currents and certain weaker large tidal currents.

Figure 4.14. *Axial flow tidal current turbine and Venturi effect*

4.3.4.3. *Wave power systems*

The ability to recover potential or kinetic energy from waves in standardized industrial conditions is underway. A large variety of concepts have been proposed, some having been demonstrated in trial basins, others at sea. Few are currently in pre-industrial phases.

The main types of technologies are categorized depending on the nature of exploitation of wave energy, and we cite here some examples of projects being developed [PAI 09]:

– exploitation of the accumulated potential energy of a water mass due to the kinetic energy of incident waves, overtopping systems, floating, e.g. WaveDragon (Denmark);

– passing through a turbine of the air trapped in a cavity subjected to internal movement of a water mass induced by the swell, systems with oscillating water columns (Figure 4.17), *onshore*, e.g. LIMPET (Scotland), and *offshore*, e.g. WaveBob (Ireland), Oceanlinx (Australia);

– utilizing the movement of water material, floating or on the seabed, caused by waves, probably the most developed concept, examples e.g. several floating bodies: Pelamis (UK, Figure 4.15), WelloOy Penguin (Finland), SEAREV (France); seabed: WaveRoller (Finland, Figure 4.18), Oyster (UK), CETO (Australia, Figure 4.16);

– exploitation of the pressure forces induced by the orbital kinematics of ocean swells.

These classifications are not set in store, and for this reason the wave power trial center European Marine Energy Centre (EMEC), in Scotland, is proposing another classification for *offshore* wave power systems based on the modes of conversion. This is presented in the Figures 4.15–4.17. All the illustrations are taken from www.aquaret.com.

Figure 4.15. *"Attenuator" type – PELAMIS P2 (© PWP 2010)*

Figure 4.16. *Punctual absorber (see color section)*

Figure 4.17. *Oscillating water column (see color section)*

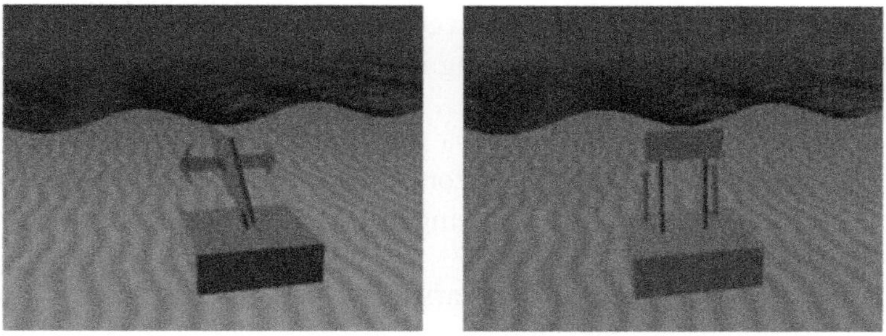

Figure 4.18. *"Horizontal" or "vertical" submerged oscillating converter (see color section)*

These different onshore and offshore systems, submerged or floating, call upon various mechanisms to convert wave energy, notably via hydraulic or all-electric chains. Unitary power goes from a few hundred kW to more than a MW. There still remain some problems of reliability of the machines tested at sea (most of the UK machines are still in the demonstration stages). The development of sea surface systems will also have to take into account the impact on navigation and on traditional coastal activities. For onshore systems, some coastal installations (dikes, quays, etc.) can also become equipped with wave energy recovering devices (for example, at Mutriku, in the Spanish Basque country, the oscillating water column technology (OWC)).

4.3.4.4. *Ocean thermal energy conversion (OTEC)*

According to Paillard *et al.* [PAI 09], the three possible uses of cold deep seawater, used in conjunction with warm surface water, are:

– the production of mechanical energy (turbine) to produce electricity (alternator) in open or closed cycles, with small temperature gaps (20°C approximately between the surface and depths of 800–1,000 m) (example of closed cycle in Figure 4.19);

– the production of freshwater from the desalination of surface seawater in processes of evaporation–condensation in chains under a very small temperature gap in stages (open cycle). The installations can be onshore or floating at sea;

– the provision of cold for direct air conditioning, an example of SWAC, with installations operating in Polynesia and Réunion.

Seawater can also be used as a thermal vector for heat pumps in coastal installations in temperate zones (industries and offices near the shore, accommodation, urban heating).

OTEC requires multi-disciplinary competences in engineering and high industrial capacities. Only two industrial projects are identifiable today, both in offshore floating platforms, one in the USA headed by Lockheed Martin, the other in France headed by DCNS. Engineering studies also exist in Japan and in the Netherlands. The offshore mode allows vertical placement above the cold source, but requires mastering the sometimes-difficult conditions of tropical seas states (cyclones). The order of magnitude of installed power is several MW to several dozen MW per floating power plant.

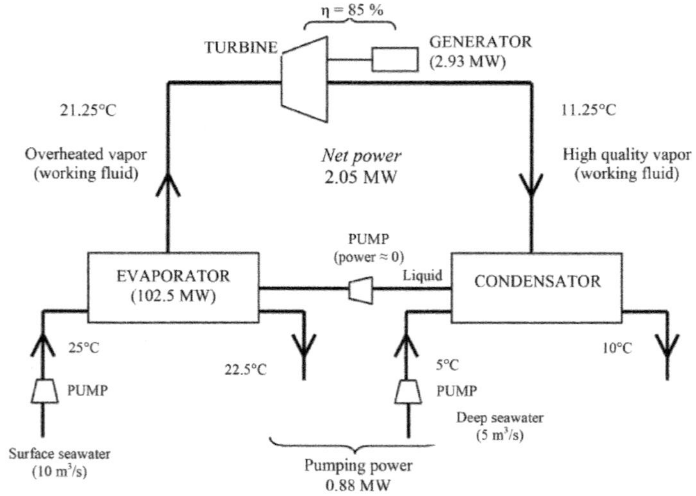

Figure 4.19. *Diagram of a typical closed cycle [NIH 11]*

The transport of cold water remains a technological issue (controlled costs and installation process on the site). Moreover, works of R&D are carried out on the onboard thermodynamic module, notably on materials, the architecture of exchangers (Figure 4.20) and the heat-transporting fluid (ammoniac or other). Heat exchangers must also be equipped with effective anti-fouling systems. The system's global yield is only of a few percentages for small temperature gap systems, and each optimization is important. But the resource is "free" and the major consequence of this small efficiency lies in the oversizing of conducts, exchangers and the turbine.

Figure 4.20. *OTEC demonstration floating module for Martinique (source DCNS) (see color section)*

4.3.4.5. *Offshore wind turbines, fixed and floating*

Fixed offshore wind turbines, set in shallow waters (10–40 m roughly), derived from terrestrial technologies, represent today the only full-scale technology in the stage of the industrial production of energy at sea. The main fields of exploitation are located in shallow seas of Northern Europe (North Sea and Baltic Sea), but also in Asia, as well as in China.

Fixed systems use different foundation technologies adapted to different depths and to the changing nature of the seabed: gravity bases, monopiles, tripods, jackets. These types of foundation were designed in the past for the petroleum industry and must be adapted to MRE. Offshore turbines are now specialized, of large dimensions (important powers: 5–7 MW, up to 8 MW in 2014), and adapted to marine conditions (for example, pressurized nacelles to avoid the introduction of salt sea air).

Figure 4.21. *Floating wind turbine: Winflo project (© Winflo) (above) and VertiWind project (© Nenuphar/Technip/EDF EN) (below)*

A major technological breakthrough is required for the installation of wind turbines in deeper waters. These are the floating systems that allow access to larger maritime zones and provide distance from crowded coasts. Several projects of development of floating wind turbine prototypes exist in France (Winflo and VertiWind projects, Figure 4.21), in Europe (a real scale demonstrator in Norway for the Hywind project headed by Statoil, and in Portugal, at a reduced scale, the Windfloat project), or in the USA and in Japan. In total, more than 20 projects exist.

Floating technologies present different options adapted to the specific conditions of the zone of exploitation (Figure 4.22):

– float type (technologies taken from offshore oil activities): semi-submersible, tension-leg platforms (TLP) or Spar (see Figure 4.22 for Hywind example);

– turbine type: axial flow (horizontal axis), used in Winflo, transverse flow (vertical axis), used in VertiWind;

– position in relation to the wind: face or back to the wind.

Figure 4.22. *Types of foundations for offshore wind turbines, left to right: pile, jacket, TLP, Spar, semi-submersible (www.aquaret.com)*

The technological challenges consist of developing, at a controlled cost, an efficient turbine requiring little maintenance and able to resist fatigue (for an operating life of around 20 years), on an anchored floater with limited movement. Electrical production must also be exported to land via a transformer substation and a cable.

4.3.4.6. *Osmotic pressure*

Not particularly well studied on a global level, this technology, currently developed in Norway (Statkraft project), uses the physical and chemical properties of salinity gradients between seawater and freshwater, with a resulting mechanical pressure able to drive a hydraulic turbine. The recovery of osmotic pressure through a semi-permeable membrane requires access on a single site to large amounts of seawater and freshwater (mouth of river, or in a fjord). The configuration of the site and the membrane technology (fragility, cost, maintenance) are the main limitations of this technology, which currently remains in experimental stages (4 kW prototype installed in Norway in 2009).

4.3.5. *Involvement in France, Europe and the world*

4.3.5.1. *Additional players that make up the sector*

The MRE sector is growing due to increased contributions of public and private agents. In France, the societal stakes involve the State and territorial collectivities, in particular regions, associations, and users. The industrial stakes involve companies.

This major incentive requires the development of new activities, where all the components of this new sector can combine. The principal agents of the research and innovation sectors are the universities and the *Grandes écoles*, research institutes, engineering centers, private R&D centers and innovating SME (small- and medium-sized enterprises). For the upstream sectors, the stakes are technological as much as they are societal, taking into account the environmental impact, the integration into existing human activities and the processes of social acceptability. For industrial development, this is up to the companies, but the State must also be closely linked with industrial political incentives (buy-back rates of kWh, public/private investment, calls for projects and offers). These large-scale projects require the mobilization of huge funds (the unit is a billion euros for offshore fixed wind farms), calling upon private and public financial bodies. Fiscal policies of incentive must also be established at the level of the State and of the European Union. Lawyers, law-makers, maritime law and energy law specialists must set the course for a stable legal context, adapted to territorial seas and the exclusive economic zone (EEZ). Planned investments must be carried out very early so that the infrastructures can be operational on time. It is the role of the State, of territorial collectivities and of companies to equip industrial and port sites, to invest in electricity distribution networks. Finally, the creation of adapted training must be anticipated, for each level of the development chain, from the specialized factory worker to the engineer.

In France and in Europe, the agents get together and form collective processes where shared interest can be pooled. Companies get together to pool support for their activities. The goal is to create an MRE lobby, and to organize the creation of this industrial sector. Professional syndicates are required, sector committees like those in France: the Syndicate for Renewable Energies (SRE), the Council for the Orientation of Research and Innovation in Shipbuilding and Related Activities (CORISRA) French Maritime Cluster (FMC), and in Europe: the Ocean Energy Association (EU-OEA), the European Wind Energy Association (EWEA). In 2012, a call for European pre-industrial projects (NER300) resulted in the co-financing, via the European Investment Bank (EIB), of several development projects of pilot farms, notably for tidal current energy or floating offshore wind energy (including the VertiMED project in France, with vertical axis wind turbines, headed by EDF Energies Nouvelles).

Local authorities are being mobilized which are initating regional initiatives, for example, in France, via regional economic, social and environmental councils (CESER in Bretagne, in Basse-Normandie and in the Pays de la Loire), studies of regional potential, calls for regional projects, and investments in port infrastructure.

4.3.5.2. *The agents and financing of R&D*

In France, the growing sector of MRE owes a lot to the scientific community, and particularly those within public research organisms taking an interest in the sea (Ifremer and *Ecole centrale de Nantes*). These researchers have demonstrated the advantages and the feasibility of extracting energy from oceans. They have mobilized entrepreneurs and public forces.

However, the MRE constitutes a system that touches on many domains of academic research, mobilizing research teams within various disciplines located throughout the country in different organisms. Thus, following the work of these pioneers, oceanograpers, fluid mechanics and energy specialists, a federal approach is needed for the emergence of a global research policy in the domain of MRE. Successively, the *Grenelle de l'environnement* and the *Grenelle de la mer*, have taken into consideration the MRE (notably through objectives of energy production and support for research).

Some realizations already carried out include:

– IPANEMA, a collective initiative that has allowed, since 2008, in France, for the federating of the industrial agents and the world of research around proposals to favor the emergence of the MRE sector;

– the *Alliance Nationale pour la Coordination de la Recherche sur l'Énergie* (national alliance for the coordination of energy research) – ANCRE – leads a programmatic approach for academic research on MRE and contributes to the defining of a national strategy (SNRI);

– the *Institut de Transition Énergétique* (ITE), whose project "*France Energies Marines*", created in 2012, finances, in public/private partnerships, works of R&D dedicated to the development of ways of conducting trials at sea. Part of the same dynamic as IPANEMA, it has over 50 members, companies, research organisms and local authorities;

– as part of a French governmental plan to develop innovation, the "*Investissements d'Avenir*" (future perspectives) (IA), the ADEME has called for several R&D projects (AMI: *Appel à Manifestation d'Intérêt*), one of them, dedicated to renewable energy systems (RESs), has been helping, since 2009, finance large-scale demonstrators for two tidal current projects, two floating offshore wind projects and for a wave power system. The AMI MRE of 2013 also co-finances R&D on technological building blocks;

– research organisms in the marine, or energy, domain develop programs on the MRE, citing Ifremer, the CNRS, the IFPEN;

– universities and *grandes écoles* have ambitious R&D and training programs on MRE, for example the *Ecole centrale de Nantes*, the *ENSTA Bretagne*, the *UBO*, the *Ecole normale supérieure de Rennes*;

– sea clusters have been labeling in the last few years just over a dozen R&D projects on technological demonstrators or equipment;

– companies invest significantly not only in R&D major industrial groups like EDF, ALSTOM, DCNS, TECHNIP or AREVA, but also in SME, like for example Nass&Wind, Sabella SAS or Nénuphar.

In Europe, the framework research and development program and its successor Horizon 2020 are also mobilized to present plans involving MRE. In the UK, the EMEC proposed trial facilities at sea for wave power systems and tidal current turbines and the National Renewable Energy Centre (NAREC) is combining trial infrastructures for offshore wind turbines. Moreover, these European projects allow shared approaches, notably citing the INTERREG projects (for example, MERiFIC between English Cornwall and the French Finistère aimed to validate site study methods), the MARINET project, which pools European test facilities and hydrodynamic basins, the ERA-NET project for the construction of European exchange networks.

The *European Energy Research Alliance* (EERA) gathers in Europe, as part of a coordination of national research efforts, most of the research laboratories on energy and includes an entity dedicated to MRE in the form of a *Joint Programme Ocean Energy*.

Multilateral relations are also in place between French, European and international agents to share R&D work, notably with the Americans (USA, Canada, Chile), the Asians (Japan, Korea) and the Australians.

International seminars dedicated to MRE are growing in size, with the example of the International Conference on Ocean Energy (ICOE) and the European Wave Energy Conference (EWEC) which is also located in Asia. Large conferences, initially dedicated to the offshore oil industry, are developing sessions dedicated to MRE (for example, the Offshore Technology Conference (OTC) in Houston, the International Conference on Ocean, Offshore and Artic Engineering (OMAE), etc.).

Today, two international institutional organisms are looking at MRE:

– the International Energy Agency (IEA), which contains a platform dedicated to MRE (OES) and notably publishes study summaries;

– the International Renewable Energy Agency (IRENA) which develops global knowledge bases on RE potentials.

In the future, the dimensions of the stakes involved in MRE must be considered, and this sector, highly demanding in innovation and R&D, needs to be considered a priority.

4.3.6. *The importance of MRE overseas and the insular context*

4.3.6.1. *Insular energetics contexts*

Insularity leads to, in most cases, a strong dependence on fossil fuels. The development of sources of RE production is a necessity for these territories. Terrestrial resources such as hydraulics, wind power, solar, geothermal energy or biomass have been the objects of development for years depending on the local context. Marine energy resources are present, to various degrees in all insular territories and, notably, overseas for France. The costs of production of the kWh from fossil fuels are such that profitability for productions using renewable energies is reached much sooner than in centralized and connected areas. Beyond French overseas territories, it is an important market for all insular territories.

4.3.6.2. *MRE resources in the various French overseas territories*

French overseas territories, with two million inhabitants and an EEZ of more the 11 million km^2, provides France with exceptional maritime dimensions. It is strongly reliant on fossil fuels, which it lacks naturally. This situation leads to the massive import of fossil fuels to satisfy its global energy requirements. More than half of these imports are for use in transport. Moreover, all of the coal used in the coal–bagasse production centers is also imported. The use of bagasse, a fibrous leftover from sugar cane, in energy production is an old practice.

The economy of current production by thermal groups in faraway islands, which is the comparative base with which to judge the relevance of alternative energy production solutions, is characterized by the integration of series of transport costs, subsidies and equalization payment funds, without mentioning environmental consequences, which is increasingly hard to support. Energetic autonomy in the production of electricity and the development of MRE is a major issue for overseas territories. However, French overseas territories are progressing toward assigned objectives (see Table 4.2), both in terms of energy efficiency and development of RE. Some territories are even exceeding their 2020 objective, such as French Guyana and Réunion, and others are on the right path: Guadeloupe, French Polynesia and New Caledonia. Martinique and Mayotte are also making good progress, but lag behind the others due to lower hydraulic and geothermal potentials.

	% RE in 2010	Objectives 2030	MRE potentials
Guadeloupe	30% (geothermal)	Autonomy	OTEC and wave SWAC
Martinique	7% (various)	Autonomy	ETM and wave SWAC
French Guyana	56% (hydraulics)	Autonomy	River resource to be evaluated
Réunion	34% (hydraulics)	Autonomy	Wave and OTEC SWAC Objective 80 to 100 MW of MRE by 2030 (GERRI)
Mayotte	8% (PV)	30% by 2030	Wave and OTEC. SWAC
French Polynesia	24.4% (hydraulics)	Autonomy 50% by 2020	OTEC and wave SWAC
New-Caledonia with Loyalty Islands	23.5% (hydraulics)	50% in 2020	Wave (Loyalty Islands) OTEC but large lagoon
Saint-Pierre and Miquelon	3% (wind)	–	No data (studies on wave and current)

Table 4.2. *Objectives in terms of energetics autonomy of electricity production and contribution of MRE in the main territories*

The potential of marine energies over all of the overseas territories is high. Their contribution to objectives of autonomy in terms of electricity by 2030 could be significant. But the estimation of these exploitable potentials has been, up to now, only the subject of punctual studies, like on Réunion. It is full-scale research subjects that need to be carried out in the various overseas territories. The *Conseil Économique, Social et Environnemental* (CESE) indicated, in 2011, that "the overseas territories have the ability to become the laboratory of RE for France and become a window for new technologies in this strategic domain".

4.3.6.3. *Examples of achievement and projects overseas*

Among these achievements, let us note the air conditioning, since 2006, of a hotel in Bora-Bora by SWAC, and more recently in Tétiaroa. Other SWAC projects are underway or being studied in French Polynesia or on Réunion, in particular for hospitals.

For OTEC, after the preliminary pilot project for a 5 MW production center in Tahiti, coordinated by Ifremer in the 1980s, several feasibility studies have been carried out since 2009 by DCNS in French Polynesia and on Réunion, as well as in Martinique where a floating production center of 10 MW is currently being studied (Figure 4.20). Studies have also been carried out by Pacific Otec in Polynesia, and others are planned for a land-based center on the Virgin Islands by DCNS and the American society, Ocean Thermal Energy Corporation.

For wave energy on Réunion, the wave power projects are in to three categories: offshore, nearshore and onshore. The prototype of the CETO concept has been in its testing phase, by ENF EN, since the end of 2012. The cyclone of January 2014 showed the need to better understand the stress suffered by structures in extreme conditions. In Tahiti, the characterization of the potential for wave energy on the South side of the island should allow for future establishment of a wave-power prototype.

Finally, most overseas territories do not possess any marine current resources, besides those from currents in atoll passes, currently being studied in Hao with the goal of building a prototypic demonstration.

4.3.6.4. *Conditions for insertion into an insular or isolated network*

The sector of MRE is an emerging one. It is still open to innovation, notably for systems adapted to the insular overseas context, including specificities in terms of biodiversity. Thus, these innovative devices must be used as demonstrators in the territories, with a follow-up in mainland R&D. New constraints are added in the insular and overseas context, like the occurrence of extreme events (cyclones) and the presence of local means for installation, exploitation and maintenance of the machines. This must lead to research into specific solutions allowing a reduction of these limitations. For MRE development to be able to contribute to the economic development of these territories, the presence of local industrial capabilities needs to be supported.

Future energy plans must include MRE in their programs by 2020 and no later than 2030, since they are expected to have achieved full capacity by then. Studies and follow-up measures are necessary (estimates of exploitable potential, evaluation of impacts, localization, regulations, fiscal matters, etc.).

Finally, the intermittent character of some of these energies, as well as of terrestrial REs, requires the association of storage solutions. Other than classical solutions, which are currently very studied, hydraulic storage projects by marine Pumped-Storage Hydroelectricity (PSH) are also being studied, notably in Guadeloupe, whose impacts will need to be evaluated.

4.4. Technological and non-technological issues

4.4.1. *Limitations of the marine environment*

4.4.1.1. *The need for detailed knowledge of environmental conditions*

The development of MRE demonstration and industrial projects at sea, established more than a dozen years ago in the case of offshore wind energy, and mainly in Europe, has shown the need to refine the characterization of the environment (waves, currents and winds) for a more accurate evaluation of the available energy resources adapted to spatial and temporal scales, but also to optimize the design of marine energy converters. The optimization of energy systems also requires specialized knowledge on the installation of the structures and their response, notably for the foundations, the interactions between the different systems and with the environment in the context of power plants and other production facilities. This knowledge is vital for power plant developers and managers, and of the electrical grid as data for their economic models.

4.4.1.2. *Minimizing the impact of hostile environments*

The design and construction of systems regularly exposed to high repetitive stress in aggressive environments requires very specific expertise. The skills gained from offshore petroleum activities is highly valued in the domain of MRE. Knowledge of the behavior and durability of materials in the marine environment must be applied, in conjuction with that of new materials with low environmental impacts, so as to optimize their use in MRE applications, especially for anchoring lines (Figure 4.23) and blades. Innovative processes and materials must still be developed in the fight against whose effects strongly influence the biofouling level of maintenance of MRE devices, and therefore the economic model of the production facilities.

4.4.1.3. *Validating technological choices in trial infrastructures and sea testing facilities*

The modeling of MRE convertor devices applies numerical tools validated by experimental studies in hydrodynamic basins and *in situ* measurements. Numerous trial basins are operational in Europe, and some of these are found in France, like at Ifremer (Figure 4.23) and at the *Ecole centrale de Nantes*. These basins allow us to evaluate systems in terms of wave, wind and current conditions. Interactions between devices must be modeled in the context of production facilities so as to study the impact of effects of different types of turbulence on both production and on the environment, notably the hydro-sedimentary effect. Structural analyses associated with resistance and fatigue of materials are also conducted using trial infrastructures (Figure 4.23) and *in situ* measurements (monitoring), like the instrumentation of tidal current turbine blades.

Moreover, sea testing facilities have been developed in Europe since the early 2000s. The pioneer being the EMEC in the Scottish Orkney Islands, which is able to test both wave power generators and tidal current turbines. In France, several sea testing facilities are currently being developed (Figure 4.24). In 2006, they were developed for the wave power site SEM-REV off the coast of Croisic, by research organisms, SME and investors with the support of collectivities and of the State. This coordination is supported by *France Energies Marines*. They are aimed at testing demonstrators but also at testing subsystems or connected systems. All the facilities will benefit from a real-time following of the environmental conditions and of the behavior of the devices, as well as a data management system. They will allow the acquisition of knowledge on the impact issues. However, the cumulative effect of these impacts cannot be studied in these facilities.

Figure 4.23. *Wind and wave trial basin and mechanical tests on synthetic cables (source: Ifremer)*

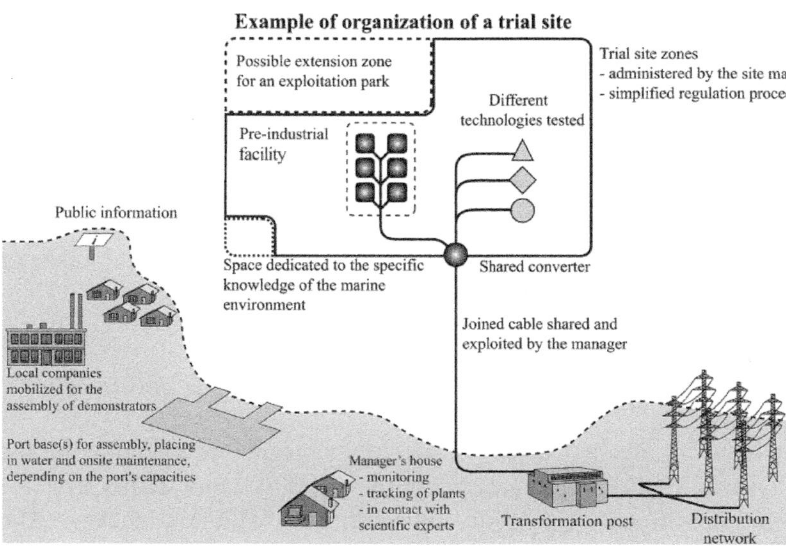

Figure 4.24. *Diagram of a typical sea test facility (adapted from Ipanema 2009)*

4.4.2. *Development of reliable and economically competitive technology*

The developers of MRE equipment and systems have the cost of the kWh in their sights. Their economic model is based on the hypothesis of grip parity, i.e. the convergence in the technologically developed stage toward the market price of electricity. Investments (including R&D), installation and exploitation costs, and risk provision must integrate this data in order to make the necessary profits for all companies. This means designing the most effective system, by adapting the following elements:

– efficiency and energy productivity: optimizing energy production;

– reliability for minimal maintenance in a highly hostile environment: minimizing exploitation costs and minimizing production disturbances;

– building systems in series at minimal cost, by limiting the requirements for specialized and costly infrastructure;

– using high performance on-site installation systems and develop dedicated maintenance and installation processes;

– controlling risks and impacts on the environment.

The equation is therefore difficult to resolve and probably only large industrial groups or consortiums are able to unite the capacity to produce these machines in series, exploit these production facilities and develop this entire sector. SME intervene during the preliminary stages of innovation, then take charge of subsystems of equipment or of services during production or exploitation.

4.4.3. *The storing of electricity, a lever for large-scale development*

MREs are mostly highly variable, but quite predictable in the short-term. With the notable exception of OTEC, production systems do not continuously provide constant and available energy. The predictability of the production is variable depending on the technology: excellent for OTEC (these are base energy resources), good for tidal current energy (predictable tidal cycles), and less precise for wind energy and wave energy.

The notion of charge factor allows for the quantification of the capacity of a system to produce at nominal power levels over a given duration. For example, OTEC could reach a charge factor of more than 80% per module (maintenance time and accidental interruptions must be considered), tidal current turbines in the order of 40% and this would fall to around 30% for offshore wind turbines.

Here we call an electricity storage system a system which allows for the transformation of electrical energy into an intermediary form of storable energy (charge), followed by its recovery (discharge), concluding with an overall acceptable yield. One of the advantages of storing energy produced is to compensate for these charge factors below 100% by introducing into the electrical system a device that increases the flexibility of production tools. Due to storage capacities, the producer can then guarantee production, for example, on the basis of meteorological predictions, to the grid manager. This storage notably avoids the use of a fossil fuel power plant when possible and helps to optimize *in fine* the management of energy to minimize the global cost over the whole of the system's lifecycle. However, we must remember that storage is just one method of flexibility among others, and that it is more costly considering the decreases of investments and losses.

Storage length is variable depending on the technology used. Short-term storage is used to improve the quality of the electricity produced and to compensate for variations of high frequency productions (in the order of the second), such as those direct from wave power technology (the production of electricity fluctuates with the rhythm of the waves). Longer-term storage allows a smoothing of the power as presented higher up and contributes to a better guarantee of production. In the case of isolated production systems (on an island for example), the latter also allows for complete autonomy of the electrical provision system.

Different technologies have been developed for industrial applications (transport, energy) and for renewable terrestrial energies;

some are not adapted to a maritime context. The reader can refer to recent publications [MUL 13].

Hydrogen is often presented as the energy vector that would allow freedom from the power cable linked to the distribution grid. The MRE device would then use its energy to carry out electrolysis of seawater to produce hydrogen and store it, but this is no longer storage (reversible) of electricity, but rather production of a combustible that is more easily transportable over long distances than electricity.

Coastal PSH projects, using seawater, help manage the intermittent nature of productions on islands (Okinawa power plant in Japan, Figure 4.25, EDF project in Guadeloupe) [LEM 11]. PSH stations at sea (artificial atolls) are also envisaged for large capacities in the order of several GWh [VIG 12].

Figure 4.25. *Marine PSH in Japan (Okinawa, 31 MW–180 MWh)*

French legislation applicable to insular systems forces producers, for penetration rates of 30% of intermittent energies, to guarantee their production in advance using better predictions and/or the use of systems. The application of marine PSH would allow rates of 50% of intermittent energy to be reached in the network.

The economic model of storage devices is struggling to impose itself; it can be competitive in isolated sites, or even in insular zones, but remains unprofitable for open grids (model of the electricity market based on non-storable electricity) in which other degrees of flexibility (hydraulic production, including land-based PSH, thermal power plants, action on consumption, etc.) are less expensive.

4.4.4. *Controlling environmental and societal impacts*

The establishment of MRE power plants could have an impact for several decades both on marine ecosystems (the physical environment and the living organisms) and on uses of the marine space (other human activities). The potential impacts of MRE must be evaluated, and then limited as much as possible so as to ensure a proper integration of future projects. Beyond the technological issues, the industrial use of marine energy is conditioned by the controlling of societal and environmental impacts, some of which are common to all MRE sectors. This condition is today a matter of consensus, but during previous studies for the installation of the tidal power plant of La Rance (in the 1960s), the evaluation of the impacts was not carried out. The follow-up, notably carried out by the *Muséum National d'Histoire Naturelle* (MNHN), showed the massive impact of the barrage, and the fragile equilibrium that has since returned to the inside of the estuary.

4.4.4.1. *Environmental impacts*

It is necessary to study the environmental impacts over all of the area of implantation of MRE projects, i.e. at the level of the power plant, but also the cable bringing the electricity produced to land (notably the zone of the land connection, which can be the most difficult). Thus, the following must be investigated:

– the environmental impact on the physical environment, i.e. the impact on the propagation of currents and waves, on hydrodynamics, on sedimentary dynamics and therefore on the nature of the substrate. These impacts must be evaluated inside MRE production facilities and in their external fields. The turbulence effects of tidal current devices, for example, through their interaction with marine depths, are likely to modify the local sedimentary dynamics, or even to generate an impact up to the coast in the case of industrial scale installation. Knowledge of the physical impacts requires the establishment of methods of numerical modeling and *in situ* measurements;

– the environmental impact on the ecosystems, meaning the modification of the structure (biodiversity) and functioning of coastal marine ecosystems (changing of the ecological role, trophic interactions). During the installation and exploitation of the power plants, the acoustic impact on certain marine species, notably marine mammals, must be studied. Solutions to avoid the propagation of noise and of vibrations during the installation phase are also studied (bubble curtains, etc.). The potential impact of cetacean migration and reproduction must also be considered. Let us note that the impacts in some cases could be considered as positive, in particular if the MRE production facilities become areas where other human activities, that are a disturbance to ecosystems, are banned. Finally, the exploitation, at a large scale, of OTEC in tropical zones leads to local changes in temperature and the composition of the water, through the rise of nutrient-rich cold water, the effects on the ecosystem of which need to be studied (artificial upwelling effect).

4.4.4.2. *Poor feedback, need for learning and follow-ups*

The impact on ecosystems of the first offshore wind farms in Europe, notably in Denmark, has been the topic of studies carried out in the domains of sedimentology, fisheries and marine biology. Some of these studies are available to the public [DEA 06]. European projects, like Equimar [EQU 11], or even works carried out as part of the IEA-OES [OES 10], have also proposed shared methods of assessment of these impacts, and some summaries of the first results. However, the ecological context is different at the local scale of French MRE projects, and if foreign feedback can be useful, multidisciplinary environmental follow-ups will need to be carried out on each facility. These studies will need to be based on acquired knowledge of the initial state of the implantation zone and will need to advance from the installation to the eventual disassembly. Methodological guides will be enriched with gradual feedback. An important stake is the knowledge of the gradually accumulated impacts with the development of the power plants along the coasts, or even of the development of other human activities at sea, so as to avoid a concentration of activities and the possible appearance of negative effects.

These studies are an integral part of the task list imposed, in France, on industrial consortiums awarded with offshore wind farm projects along the French coast. Follow-up and consultation authorities, combining all of the parties involved (6 facilities for 3,000 MW from 2020), are also organized around each site by prefects and maritime prefects so as to avoid the insertion of this new sea-based activity becoming a source of potential conflicts. It is reasonable to assume that with the gradual improvement of our knowledge of the impacts, and the solutions to reduce these, acceptability will increase. This is even truer since these sectors will create local jobs.

4.4.4.3. *The acceptability of MRE by joint planning*

The conditions of installation and exploitation of MRE production facilities must be accepted by the population and users of the sea. Users are not only fishermen, but also agents in a whole range of activities, including maritime transportation, leisure, tourism, etc. In France, calls for offers for MRE production facilities (offshore wind energy currently, and then pilot tidal power plants) are preceded by a planning phase, coordinated by maritime prefectures and regional prefectures, with the goal of defining the installation zones of future MRE production facilities at sea, in conjunction with all of the users and associations. For offshore wind energy, two zoning phases have been carried out to define six installations zones of offshore wind farms for around 2018 and 2020. A similar planning process was undertaken in 2013 for the development of the tidal current turbine sector in France. Several sites have been identified off Basse-Normandie and Bretagne.

The exploitation of different energy sources, fossil or renewable, presents an impact that is more or less significant on biodiversity. Energy choices must be made with a global vision of what is at stake. For MRE, as was the case in Germany for the development of offshore wind energy, it is a strategic approach on the long-term that needs to be developed.

4.4.5. *Establishing an adapted regulatory context*

The creation of MRE production facilities meets a challenge when facing regulations that are not well-adapted, although constantly changing. Adaptations and simplifications of the legislation and regulations in line with the MRE domain are expected. With this development of both fixed offshore wind power and tidal current power, as announced by the government, guarantees are needed to ensure the safety of investments, notably on the visibility of the sector in France in the long-term and on the mechanism of financial support available for the development of these technologies.

Currently, three administrative authorizations are required for the development of a MRE project in French waters: (1) authorization for the occupations of a public maritime domain, (2) authorization concerning the law on waters, (3) authorization for exploitation. The need for a building permit for the construction of offshore wind farms was removed in the law Grenelle 2, but a permit is still needed for the construction of land-based equipment. The establishment of a unique office, centralizing the procedures, will allow simplifying the processes and speeding up access to authorizations. Finally, there is still no regulation for the installation of MRE production facilities outside of national waters or overseas. With the large-scale development of some technologies, like floating offshore wind turbines, EEZ regulation is expected.

It is the mechanism of calls for offers that has been retained for the industrial development of fixed offshore wind farms in France. Authorization for exploitation is granted to the winner and a buying contract for their production, at a rate determined in the offer, is written up. Outside of this mechanism, a regulated rate is applied. Installations that use wave energy, tidal energy or hydrokinetic energy are concerned by the decree on tariffs of the 1st March 2007, fixing the conditions for the buying of electricity produced at 150 €/MWh. This tariff is not adapted. The UK, Portugal and other countries have put in place mechanisms that are more attractive so as to favor the emergence of undeveloped sectors. Indeed, a tariff must take into consideration that the development of these sectors passes through the creation of demonstrators in sea test facilities, and then pilot facilities, before the exploitation of full-scale industrial power plants.

4.5. Socio-economic consequences

4.5.1. *The industrial stakes*

4.5.1.1. *Introduction*

The development of MRE must contribute to two main stakes that affect the global economy of our societies: the energetic stakes and the industrial stakes. MREs will contribute to supplying energy, but must also allow for the development of important job-creating economical activities. These are major industrial stakes and major stakes in the creation of a new sector, based on complementary domains, within companies interacting with agents of research, territorial and State agents.

R&D is highly called upon for this sector of innovation; the domains concerned are large. For example, technological: the evaluation of the energy resource, the behavior of materials in the marine environment, the storage of energy at sea, but also non-technological: the impact of MRE systems on fauna, flora or on existing human activities must be studied. Engineering allows the conception of reliable technologies, efficient and at a reasonable cost. The production in series of full-scale systems and of equipment requires the conception and construction of production lines, preferably in coastal regions. The phases of installation, exploitation and maintenance, before disassembly, allow for the production of systems with lifetimes of more than 20 years at sea. For large production facilities, accommodation needs to be built at sea, close by to host the maintenance and intervention teams.

The MRE sector needs to develop a large body of support to access these installations at sea, so as to participate in maintenance and in offshore installations. The phases of construction and of assembly of structures at ports will mobilize most existing port agents. Rapid vessels adapted for access to offshore systems are designed, built and exploited by port and maritime companies. The establishment of training programs must be planned to prepare the recruitment of competent staff for all of the components of the sector.

France's strengths in this respect are significant:

– presence of an industry with large complex systems in the following sectors: naval, offshore petroleum, energy and environment;

– presence of SME and SOHO (small office/home office) providing innovation;

– the presence of internationally renowned scientific teams and high capacity trial equipment, particularly in the sciences and marine technologies;

– port capacities to be exploited;

– a central geographical maritime position in Europe;

– a massive exclusive economic maritime zone (the 2nd largest in the world, thanks to overseas territories).

The development of these sectors, whose activities are difficult to relocate abroad, must mobilize all agents: institutional (at a European, national and regional level), industrial (groups and SME), scientific (valorization of research), of training organisms (at all levels), investors and financial bodies, port and maritime agents, operators of distribution and energy networks.

4.5.1.2. *From design to disassembly*

A full-scale sector needs to be established. The most effective approach in the long-term is to consider the development of this activity as global: by the establishment of principles of sustainable development using lifecyle analysis (LCA) .

All of the resources and transformations applied over a system's life must be considered (energy and materials used, CO_2 emissions, water consumption, production equipment, maintenance, human resources, impact on the environment and interactions with other coastal human activity, etc.). All the development phases must be considered: design and engineering, production, installation, maintenance and exploitation, and disassembly.

Each of these phases provides weighted data and can bring validated and numbered elements that allow us to have a global image of the costs and profits of the sector for society. The LCA approaches, difficult to execute from an operational point of view, exist in part for other industrial sectors (automobile, electronic goods, aeronautical and naval), and still need to be researched and experimented on further before the desired generalized application can happen.

The lifecycle of a typical project of the development of MRE technology, from conception in series production, is graded in the industry on the Technology Readiness Level (TRL) scale. This standardized scale, established in the USA, qualifies the levels of development for each stage of development.

Table 4.3 presents this classification.

Level	Characteristic	Example
1	Basic principle observed and reported	Academic research
2	Concepts and/or applications of the formulated technology	Academic research
3	Tested and analyzed critical function and/or characteristic proof of the concept	Upstream studies
4	Laboratory validation of the component and/or of the artifact produced	Numerical model, engineering
5	Validation in a significant environment of the component and/or the artifact produced	Test equipment in basin
6	Demonstration of the system/sub-system model and of the prototype in a significant environment	Test system in basin
7	Demonstration of the prototype system in an operational environment	Demonstrator at sea
8	Complete real system and qualification through tests and demonstrations	Pre-industrial prototype
9	Real system proven through successful operations/missions	Production in series

Table 4.3. *Technology Readiness Level (TRL)*

For example, the industrial ranking for fixed offshore wind turbine is now at level 9 and level 7 to 8 for tidal current turbines and some floating offshore wind projects. Beyond level 9, in the installation phases, exploitation and maintenance only concern fixed offshore wind turbines. The disassembly phases have not yet been evaluated. The feedback from fixed offshore wind turbines in Northern Europe already allows for the identification of some developmental issues for the industrial phases, in particular:

– the installation systems of power plants and, notably, specialized vessels, rare and therefore expensive;

– port infrastructure and seaside infrastructure, often saturated by existing coastal activities;

– training programs not always adept to provide a qualified and adapted workforce;

– acceptability and regulatory aspects: the process between the decision to develop a site and the start of exploitation is long (around ten years) ;

– production capacities and supply (primary materials, equipment) can also become complicated if the projects multiply at the same time.

Investment and production costs are often calculated from the CAPEX (initial investment: construction to installation at sea) and the OPEX (exploitation, maintenance and disassembly).

Depending on the technologies, their level of development and the installation zones, we provide the following evaluations (from Indicta, 2014):

– fixed offshore wind energy: investment cost in the order of 3 to 3.5 M€/MW, or a complete production cost in the order of 150 to 250 €/MWh currently, depending on the zones (North Sea, France, etc.), but in decline;

– tidal current energy: investment cost of around 3.5 M€/MW from the first 1,000 units produced, or a production cost of 150 to 350 €/MWh depending on the zones and the development of the devices (UK, France, etc.), with an objective cost of 100 to 130 €/MWh before 2030;

– floating offshore wind energy: a starting investment cost of 3 to 4 M€/MW, or a production cost in the order of 200 to 250 €/MWh, the floating turbines could reach production costs in the order of 100 to 130 €/MWh by 2030;

– wave energy: a cost for the MW estimated between 4 and 5 M€ at the beginning, and then 3–4 M€/MW from the first 1,000 units produced, with a goal of 100 to 130 €/MWh by 2030;

– OTEC: an investment cost of 20 M€/MW for the first commercial systems, or a production cost of around 350 to 400 €/MWh (still competitive for isolated territories), ultimately: 200–300 €/MWh.

As a reminder: land-based wind farms in Europe currently present production costs that are below 80 €/MWh with high charge factor machines.

4.5.1.3. *A job-creating sector*

The MRE sector (or sectors, if we consider the different technologies) generates important economic activity, providing a large amount of jobs in numerous sectors: equipment industry, development of exploitation sites at sea, energy production and distribution, materials industry (metals, composites and cements), port and maritime logistics, sea operations, environmental engineering, non-technological services (financial, insurance and legal).

The current model of fixed offshore wind energy in Northern Europe is the only one that is currently providing significant feedback. In 2012, the EWEA accounted for 35,000 direct jobs to an accumulated 5 GW. This model shows a massive localization of jobs close to exploitation sites and predicts the creation of 300,000 direct and indirect jobs in 2020.

The maritime sector is at the front line of these newly created jobs (naval and offshore, notably). Most of the time, these are existing jobs in the energy and maritime sectors. All job levels are involved, from the specialized factory worker to the engineer. There are very few new types of jobs, but rather the mobilization of competences already existing within other sectors like the naval sector, but also the offshore petroleum sector, the energy sector and service jobs (for example, environmental studies, engineering, legal workers, sailors). In 2013, the *Groupement des Industries de Construction et d'Activités Navales* (Grouping of construction industries and naval activities) – GICAN – has evaluated that 600 direct jobs of the maritime sector in France are already mobilized by the MRE sector.

In terms of employment, the impact in France and in Europe will be all the greater as implemented industrial policies will favor the installation of decision centers, of R&D and of production factories for elements with high added value, like turbines or complex equipment.

4.5.2. *Impacts and usage conflicts to control – compensatory measures*

MREs will need to show that they can be integrated in the sector from an environmental point of view, that they do not disturb pre-existing human activity and that they are beneficial to economic activity, at a local, regional and national level. With the development of all MREs sectors, follow-ups of the professional and recreational activities associated with a methodology for the assessment of ecological and societal impacts must be developed so as to know which type of compensatory measures must be used to compensate for possible losses of ecosystemic services (services of ecological regulation, collection services and cultural services). It involves evaluating the costs associated with these measures and the effectiveness of these ecologically, economically and socially. Currently, for fixed offshore wind energy projects, a tax is collected by the *Fonds national de compensation de l'énergie éolienne en mer* (national compensation fund for offshore wind energy), distributed, as financial compensation, at 50% for coastal communities from which the installations will be visible, at 35% to the national committee of members of professions whose activities are linked to fishing and marine harvesting, and at 15% for the financing of projects relating to sustainable development of other maritime activities.

4.6. Perspectives

4.6.1. *Development of the different sectors: present, short- and long-term*

Several marine resources should contribute significantly to the global energy mix in the medium- and long-term (Figure 4.9). Offshore wind energy systems (floating and fixed) and tidal current energy systems are those that, due to their high level of development, have the greatest potential for industrial development in the coming decade. However, the more complex development of wave energy systems should take place in the medium- to long-term (after 2020). The development of an OTEC demonstrator, despite being a base energy sector, is hindered by its investment costs, which are currently high, requiring institutional support to initiate the onset of the sector. It would then develop in the medium-term, in function of the evolution of the energy context, in inter-tropical zones. However, the use of cold deep water for air conditioning will develop, notably, in new installations, allowing for the creation of cold water networks, or even the valorization of this water rich in nutritional salts.

4.6.2. *Today in the world, in Europe and in France*

At the start of 2014, the global installed offshore wind capacity was of 7,100 MW, with around 500 MW in China and 6,500 MW in Europe, mainly in the UK or in Denmark. For offshore wind energy projects, 6,000 MW were in their installation phase, and 22,000 MW had received building authorizations.

In 2011, the IEA stated that the global installed MRE capacity (not including offshore wind turbines) was of 519 MW, mainly corresponding to tidal power productions, including the one at La Rance. An increase the number of industrial projects in the domain of tidal power, are expected in the years to come, in Europe but also in North America and in Asia.

However, following calls for offers in July 2011, four offshore wind energy projects were retained, for a total installed power of 1,910 MW before 2020: Fécamp (480 MW), Courseulles-sur-Mer (450 MW), Saint-Nazaire (450 MW) and Saint-Brieuc (500 MW). A second call for tender, made in spring 2013, concerns two zones: Tréport (between 480 and 500 MW) and the Ile d'Yeu and the Ile de Noirmoutier (480–500 MW). The cumulated installed power after 2020, in the order of 3,000 MW, will not satisfy the objectives of the *Grenelle de l'environnement* and the *Grenelle de la mer*.

With regards to other energy resources, the AMI ADEME in 2009 led to the co-financing of several demonstrators in the sectors of: tidal current energy, wave energy and floating offshore wind energy. The state of advancement of the projects and the difficulties met by some projects, led us to believe that the first sea installations would be coming into existence from 2013. The AMI of 2013 combines the overcoming of technological issues of technological building blocks and of demonstrators. A call for tender for several piot tidal power plant, preceded by a preliminary phase of discussion and consultation aimed at identifying the zones concerned by the call for tender, start off the development of the sector. The RTE (*Réseau de Transport de l'Électricité*, or the electricity transport network) conducted in 2012 a study aimed at proposing methods for evacuating the production of tidal power plants off Cotentin, where the resource is the greatest in the whole of France. Finally, a methodological study on the environmental and socio-economic impacts of MRE was published in 2012, and in turn, a study guide of the impact for offshore wind farms in 2013 (DGEC).

A study on the "state of the art and propositions on marine energy", given to the CGEDD and to the CGEIET by the ministers of Ecology and Sustainable Development, of Productive Revival and to the delegate minister for Marine Transport and Fishing, is one of the issues brought to the National Debate on Energy Transition of 2013.

4.6.3. *2015–2025: from demonstrators to power plants via pilot facilities*

Beyond the pursuit of support for demonstrators, for some sectors, the demonstration of the technical and economical reliability of a production facility, made up of a few machines, in real exploitation conditions at pilot sites has been started.

In France, the first pilot site should be up and running in 2015, off the coast of Paimpol-Bréhat. It is a tidal farm with four 500 kWh tidal current turbines using Open Hydro technology, placed at 40 m of depth on gravity tripod foundations. The project is headed by EDF.

Other pilot sites, of tidal current turbines and floating offshore wind turbines, should be operational before the next decade in Europe.

Several initiatives for wave energy are being developed, notably in Scotland, off the coasts of Cornwall, and in Portugal and Spain.

Pilot sites	Technology			
	Tidal current	*Floating wind*	*Wave*	*OTEC*
Number of machines per pilot farm	3–10	5–15	5–10	1
Power per pilot farm	3–17 MW	20–40 MW	5–10 MW	Around 5 MW
Total volume of the required pilot farms (all agents)	Around 50 MW	Around 70 MW	Around 30 MW	Around 5 MW

Table 4.4. *Technical parameters of pilot sites according to the SER (2013)*

Figure 4.26. *Axial flow OpenHydro tidal current turbine on its installation barge in Brest (source: EDF/DCNS, 2011)*

The industrial development of tidal current energy should be on track for some technological progress before 2020, and later for wave energy. However, according to the European Association of Marine Energy (EU-OEA), 3.6 GW of MRE (not including offshore wind energy) should be installed by 2020 [OEA 10]. This estimation, in light of what has actually been installed in 2014, does not seem very realistic.

Moreover, the industrial sector of offshore wind energy is already very active, with ambitious objectives, notably in Germany.

According to the EWEA, in 2020 the installed capacity of offshore wind energy systems should be in the order of 40 GW, or the equivalent of 4% of the electricity demand (148 TWh annual production). This estimation also seems unrealistic.

4.6.4. *2030 and beyond: large-scale development*

According to the EWEA, for offshore wind energy systems, a capacity of 150 GW should be operational by 2030 for an annual production of 562 TWh covering 14% of European demand of electricity, reducing CO_2 emissions by 315 million tons. 60% of this capacity would be installed in the North Sea. These projections seem optimistic as this gain of 110 GW over 10 years would require the installation of more than 900 MW monthly in Europe over 10 years. On top of the financial issues, several factors can interfere, such as pressure on primary materials or the difficulty for companies to produce, but also insufficient dedicated vessels for installation and maintenance. In terms of employment, these projections would lead to the creation of 170,000 FTE (full-time equivalent) positions in 2020 and 300,000 FTE in 2030. The development on this scale of the sector will only be possible with coordinated and agreed planning in Europe, also in terms of electrical grids, which gives visibility to investors. Indeed, the cumulated investments between 2021 and 2030 would be in the order of close to 150 billion euros. However, it is probable that the floating offshore wind energy sector will have a prominent position, significantly greater than that of the fixed offshore wind energy, as soon as the sector reaches technical and economic viability.

Europe could also be in the front row for MRE, with the European Commission roadmap planning an installed capacity of 188 GW in 2050 (not including fixed and floating offshore wind farms), which would represent 15% of the European energy mix and an annual market of 15 billion euros. According to the OES, the MRE (not including offshore wind energy) also represents a considerable source of employment, and would lead to the creation of 160,000 direct jobs between now and 2030, and more than 450,000 direct and indirect jobs between now and 2050 [ERN 12].

According to a recent study carried out by Indicta for the *Groupement des industries de construction et activités navales* (Grouping of construction industries and naval activities) – GICAN [GIC 13] – in France (both mainland and overseas), all of the MRE (including offshore wind energy) would represent 15 GW of installed power in a median scenario (or 50,000 direct jobs in France) and 20 GW, in a more optimistic version, in 2030. According to these scenarios, by 2030, fixed offshore wind installed capacity would reach 6 GW, tidal current installed capacity would represent 3 GW and wave energy systems associated with OTEC at an installed capacity of 2 GW. Floating offshore wind energy adjust the total in function of the two options, with the average scenario placing them at 4 GW and the optimistic scenario at 9 GW. However, the global installed capacity in 2030 should not have reached 100 GW.

Thus, the MRE (not including fixed offshore wind energy) represent one of the rare domains of RE in which France can claim a strong place among the leaders. But to solidify this ambition, large investments in R&D, with the support of programs such as IA (*Investissements d'Avenir* or future investments), are required. In the medium-term, the, or future investments, large-scale use of MRE depends on the capacity of those in the sector to deal with the numerous technological and economic challenges and to affirm themselves in the face of competition from international rivals.

4.7. Bibliography

[CHA 97] CHABOT B., *Energies renouvelables*, Encyclopedia Universalis, 1997.

[DEA 06] DEA (DANISH ENERGY AGENCY), Environmental impact, Offshore wind power, 2006. Available at www.ens.dk/en-us/supply/renewable-energy/windpower/offshore-wind-power/environmental-impacts/sider/for side.aspx.

[ERN 12] ERNST & YOUNG, Les énergies marines renouvelables: quelles opportunités pour la France, Bordeaux, 2012.

[EQU 11] EQUIMAR, Protocols for the Equitable Assessment of Marine Energy Converters, 2011. Available at www.eu-oea.com/projects/ equimar-3/.

[GIC 13] GICAN GROUPEMENT DES INDUSTRIES DE CONSTRUCTION ET ACTIVITÉS NAVALES), Potentiel de création d'emplois et de valeurs lié aux EMR, Press Report, 2013.

[IEA 13] IEA (INTERNATIONAL ENERGY AGENCY), Key World Energy Statistics, 2013.

[IES 12] IES (INTERNATIONAL ENERGY AGENCY), An International Vision for Ocean Energy, Ocean Energy Systems Implementing agreement, October 2012.

[IPC 07] IPCC, Climate Change 2007: Synthesis Report, An Assessment of the Intergovernmental Panel on Climate Change, 2007.

[IPC 12] IPCC, Special Report of the Intergovernmental Panel on Climate Change (IPCC) Renewable Energy Sources and Climate Change Mitigation, 2012.

[LEM 11] LEMPERIERE F., Les stations de transfert d'énergie par pompage, 2011. Available at www.hydrocoop.org/fr/publications/Stockage-d-energie-par-pompage-d-eau-de-mer-01.2012.pdf.

[MUL 09] MULTON B., CLÉMENT A.H., RUELLAN M., *et al.*, "Marine Energy Resources Conversion System", in SABONNADIERE J.C. (ed.), *Renewable Energy Technologies*, Chapter 7, ISTE, London and John Wiley & Sons, New York, pp. 181–226, 2009.

[MUL 11] MULTON B., THIAUX Y., BEN AHMED H., "Consommation d'énergie, ressources énergétiques et place de l'électricité", *Techniques de l'ingénieur,* Traité de génie électrique, D3900v2, February 2011.

[MUL 12] MULTON B. (ed.), *Marine Renewable Energy Handbook*, ISTE, London and John Wiley & Sons, New York, pp. xvi–xx, 2012.

[MUL 13] MULTON B., AUBRY J., HAESSIG P., *et al.*, "Systèmes de stockage d'énergie électrique", *Techniques de l'Ingénieur*, BE8100, 2013.

[NIH 12] NIHOUS G., GAUTHIER M., "Ocean thermal energy conversion: A historical perspective", in MULTON B., (ed.), *Marine Renewable Energy Handbook,* Chapter 12, ISTE, London and John Wiley & Sons, New York, pp. 367–404, 2012.

[OBS 13] OBSERV'ER, La production d'électricité d'origine renouvelable dans le monde, 14th ed., 2013.

[OEA 10] OEA, Roadmap des énergies marines en Europe, Oceans of Energy 2010-2050, EU-OEA, 2010.

[OES 10] OES-IA, Annex IV: Environmental Effects of Marine and Hydrokinetic Devices, 2010. Available at www.ocean-energy-systems.org/oes_documents/annex_iv_reports/environmental_effects_of_marine_and_hydrokinetic_devices.

[PAI 09] PAILLARD M., LACROIX D., LAMBLIN V. (eds.), *Marine renewable energies – Prospective foresight study for 2030*, QUAE, vol. 5, 2009.

[SAL 12] SALOMON T., JEDLICZKA M., MARIGNAC Y., *Manifeste négaWatt*, Actes Sud, Arles, 2012.

[SAT 11] SATHAYE J., LUCON O., RAHMAN A., "Renewable energy in the context of sustainable development", *Special Report on Renewable Energy Sources and Climate Change Mitigation*, Chapter 9, IPCC, 2011.

[SOS 13] SOS (SERVICE DE L'OBSERVATION ET DES STATISTIQUES), Chiffres-clés du climat France et Monde, ministère de l'Ecologie, du Développement durable et de l'Energie, 2013.

[VIG 12] VIGNY J.P., Les STEP marines et leur avenir, Université d'été SLC, Aber Wrac'h, september 2012.

[WEC 10] WEC (WORLD ENERGY COUNCIL), Survey of Energy Resources, 2010.

List of Authors

Jean-Baptiste BÉRARD
Ifremer – LBPA
Nantes
France

Marc BŒUF
DCNS – IEMR
Brest
France

Gaël BOUGARAN
Ifremer – LBPA
Nantes
France

Jean-Paul CADORET
Ifremer – LBPA
Nantes
France

Grégory CARRIER
Ifremer – LBPA
Nantes
France

Aurélie CHARRIER
Ifremer – LBPA
Nantes
France

Noémie COULOMBIER
Pôle Marin – ADECAL
Nouméa
New Caledonia
France

Cécile DEBITUS
IRD – UMR EIO
Papeete
French Polynesia

Daniel FAGET
MMSH UMR Telemme
Aix-en-Provence
France

Matthieu GARNIER
Ifremer – LBPA
Nantes
France

Jean GUÉZENNEC
Ifremer – BM
Brest
France

Raymond KAAS
Ifremer – LBPA
Nantes
France

Loïc LE DÉAN
Ifremer – LBPA
Nantes
France

Ewa LUKOMSKA
Ifremer – LBPA
Nantes
France

André MARIOTTI
University Pierre and Marie Curie
Paris
France

André MONACO
Cefrem/CNRS – UPVD
Perpignan
France

Bernard MULTON
CNRS – SATIE
Bruz
France

Elodie NICOLAU
Ifremer – LBPA
Nantes
France

Michel PAILLARD
Ifremer – LCSM
Brest
France

Jean-CHARLES POMEROL
University Pierre and Marie Curie
INSIS/CNRS
Paris
France

Patrick PROUZET
Ifremer – DS
Issy-les-Moulineaux
France

Catherine ROUXEL
Ifremer – LBPA
Nantes
France

Jacques SACCHI
Ifremer – DHMT
Sète
France

Bruno SAINT-JEAN
Ifremer – LBPA
Nantes
France

Nathalie SCHREIBER
Ifremer – LBPA
Nantes
France

Index

Other titles from

ISTE

in

Oceanography and Marine Biology

2014

Seas and Oceans Set coordinated by MONACO André, PROUZET Patrick
Ocean in the Earth System
The Land-Sea Interactions

MUTTIN Frédéric
Marine Coastal and Water Pollutions: Oil Spill Studies

2013

LÉCUYER Christophe
Water on Earth: Physicochemical and Biological Properties

Printed and bound by CPI Group (UK) Ltd, Croydon, CR0 4YY

13/09/2026

14961077-0001